T0224512

Communications in Computer and Information Science   **1169**

Commenced Publication in 2007
Founding and Former Series Editors:
Phoebe Chen, Alfredo Cuzzocrea, Xiaoyong Du, Orhun Kara, Ting Liu,
Krishna M. Sivalingam, Dominik Ślęzak, Takashi Washio, Xiaokang Yang,
and Junsong Yuan

More information about this series at http://www.springer.com/series/7899

Quan Yu (Ed.)

Space Information Networks

4th International Conference, SINC 2019
Wuzhen, China, September 19–20, 2019
Revised Selected Papers

 Springer

Editor
Quan Yu
Institute of China Electronic Equipment
Beijing, China

ISSN 1865-0929   ISSN 1865-0937   (electronic)
Communications in Computer and Information Science
ISBN 978-981-15-3441-6   ISBN 978-981-15-3442-3   (eBook)
https://doi.org/10.1007/978-981-15-3442-3

This Springer imprint is published by the registered company Springer Nature Singapore Pte Ltd.
The registered company address is: 152 Beach Road, #21-01/04 Gateway East, Singapore 189721, Singapore

Preface

This book collects the papers presented at the 4th Space Information Network Conference (SINC 2019), an annual conference organized by the Department of Information Science, National Natural Science Foundation of China. SINC is supported by the key research project of the basic theory and key technology of space information network of the National Natural Science Foundation of China, and organized by the "space information network" major research program guidance group. The aim is to explore new progress and developments in space information networks and related fields, to show the latest technology and academic achievements in space information networks, to build an academic exchange platform for researchers at home and abroad working on space information networks and industry sectors, to share the achievements and experience of research and applications, and to discuss the new theory and new technologies in space information networks. There were two tracks in SINC 2019: Architecture and Efficient Networking Mechanism and Theories and Methods of High Speed Transmission.

This year, SINC received 118 submissions, including 83 English papers and 35 Chinese papers. After a thorough reviewing process, 23 outstanding English papers were selected for this volume (retrieved by EI), accounting for 27.7% of the total number of English papers.

The high-quality program would not have been possible without the authors who chose SINC 2019 as a venue for their publications. We are also very grateful to the Academic Committee and Organizing Committee members, who put a tremendous amount of effort into soliciting and selecting research papers with a balance of high quality, new ideas, and new applications.

We hope that you enjoy reading and benefit from the proceedings of SINC 2019.

November 2019   Quan Yu

Organization

SINC 2019 was organized by the Department of Information Science, National Natural Science Foundation of China; the Department of Information and Electronic Engineering, Chinese Academy of Engineering; China InfoCom Media Group; and the *Journal of Communications and Information Networks*.

Organizing Committee

General Chairs

Quan Yu	Institute of China Electronic Equipment System Engineering Corporation, China
Jianya Gong	Wuhan University, China
Jianhua Lu	Tsinghua University, China

Steering Committee

Zhixin Zhou	Beijing Institute of Remote Sensing Information, China
Hsiao-Hwa Chen	National Cheng Kung University, Taiwan, China
George K. Karagiannidis	Aristotle University of Thessaloniki, Greece
Xiaohu You	Southeast University, China
Dongjin Wang	University of Science and Technology of China, China
Jun Zhang	Beihang University, China
Haitao Wu	Chinese Academy of Sciences, China
Jianwei Liu	Beihang University, China
Zhaotian Zhang	National Nature Science Foundation of China, China
Xiaoyun Xiong	National Nature Science Foundation of China, China
Zhaohui Son	National Nature Science Foundation of China, China
Ning Ge	Tsinghua University, China
Feng Liu	Beihang University, China
Mi Wang	Wuhan University, China
ChangWen Chen	The State University of New York at Buffalo, USA
Ronghong Jin	Shanghai Jiao Tong University, China

Technical Program Committee

Jian Yan	Tsinghua University, China
Min Sheng	Xidian University, China
Junfeng Wang	Sichuan University, China
Depeng Jin	Tsinghua University, China
Hongyan Li	Xidian University, China
Qinyu Zhang	Harbin Institute of Technology, China
Qingyang Song	Northeastern University, China
Lixiang Liu	Chinese Academy of Sciences, China

Weidong Wang	Beijing University of Posts and Telecommunications, China
Chundong She	Beijing University of Posts and Telecommunications, China
Zhihua Yang	Harbin Institute of Technology, China
Minjian Zhao	Zhejiang University, China
Yong Ren	Tsinghua University, China
Yingkui Gong	University of Chinese Academy of Sciences, China
Xianbin Cao	Beihang University, China
Chengsheng Pan	Dalian University, China
Shuyuan Yang	Xidian University, China
Xiaoming Tao	Tsinghua University, China

Organizing Committee

Chunhong Pan	Chinese Academy of Sciences, China
Yafeng Zhan	Tsinghua University, China
Liuguo Yin	Tsinghua University, China
Jinho Choi	Gwangju Institute of Science and Technology, South Korea
Yuguang Fang	University of Florida, USA
Lajos Hanzo	University of Southampton, UK
Jianhua He	Aston University, UK
Y. Thomas Hou	Virginia Polytechnic Institute and State University, USA
Ahmed Kamal	Iowa State University, USA
Nei Kato	Tohoku University, Japan
Geoffrey Ye Li	Georgia Institute of Technology, USA
Jiandong Li	Xidian University, China
Shaoqian Li	University of Electronic Science and Technology of China, China
Jianfeng Ma	Xidian University, China
Xiao Ma	Sun Yat-sen University, China
Shiwen Mao	Auburn University, USA
Luoming Meng	Beijing University of Posts and Telecommunications, China
Joseph Mitola	Stevens Institute of Technology, USA
Sherman Shen	University of Waterloo, Canada
Zhongxiang Shen	Nanyang Technological University, Singapore
William Shieh	University of Melbourne, Australia
Meixia Tao	Shanghai Jiao Tong University, China
Xinbing Wang	Shanghai Jiao Tong University, China
Feng Wu	University of Science and Technology of China, China
Jianping Wu	Tsinghua University, China
Xiang-Gen Xia	University of Delaware, USA
Hongke Zhang	Beijing Jiaotong University, China

Youping Zhao	Beijing Jiaotong University, China
Hongbo Zhu	Nanjing University of Posts and Telecommunications, China
Weiping Zhu	Concordia University, Canada
Lin Bai	Beihang University, China
Shaohua Yu	FiberHome Technologies Group, China
Honggang Zhang	Zhejiang University, China
Shaoqiu Xiao	University of Electronic Science and Technology of China, China

Contents

Theories and Methods of High Speed Transmission

Architecture and Efficient Networking Mechanism

Throughput Evaluation and Ground Station Planning for LEO Satellite Constellation Networks

Shuaijun Liu[1(✉)], Tong Wu[2], Yuemei Hu[1], Yichen Xiao[1], Dapeng Wang[1], and Lixiang Liu[1]

[1] Institute of Software Chinese Academy of Sciences (ISCAS), Beijing, China
shuaijun@iscas.ac.cn
[2] China National Institute of Metrology (NIM), Beijing, China

Abstract. With the development of satellite networks, new generation of low earth orbit (LEO) satellite constellation network composed of mega satellites has emerged. The ground stations (GS) deployment has the key impact on the system throughput for LEO satellite constellation network. However, the GS deployment planning is challenging due to the spatial and temporal distribution of traffic demands and the time-varying topology of the LEO satellites. To solve this problem, an iterative GS deployment based on marginal revenue maximization (IGSD-MRM) is proposed. The key idea is to select the geographical location, which can achieve the maximum marginal revenue, as current GS deployment location in each iteration. The simulation results show the effectiveness of the proposed IGSD-MRM algorithm in throughput improvement under different GSs numbers and feeder link antennas.

Keywords: LEO satellite network · Ground station deployment · System throughput · Iterative method

1 Introduction

With the rapid development of LEO satellite communication networks and on-board processing (OBP) technologies, the new generation of LEO satellite constellation network composed of mega LEO satellites such as OneWeb has emerged and become a development trend [1, 2]. Compared to geostationary earth orbit (GEO) satellites, LEO satellites have smaller propagation loss and end-to-end delay, facilitating user terminals (UTs) miniaturization and real-time service transmission. Compared to the Iridium and Globalstar systems appeared in the 1990s, the new generation of LEO constellation network has more advanced OBP capabilities, higher-order modulation schemes, lower manufacturing and launch cost [3, 4]. To provide the worldwide service, distributed ground stations are needed to implement the network management and data switching operation. To further improve the system throughput, the GS needs to fully consider

This work is supported by the project "Study on the Inter-satellite Networking and Data Sharing Technologies," National Key Research and Development Plan 2016YFB0501104.

© Springer Nature Singapore Pte Ltd. 2020
Q. Yu (Ed.): SINC 2019, CCIS 1169, pp. 3–15, 2020.
https://doi.org/10.1007/978-981-15-3442-3_1

the characteristics of satellite constellation operation, traffic demands and wireless link quality. How to evaluate the system throughput and make the ground station planning has key meaning and motivates this paper.

Many researchers have made works on throughput evaluation and ground station planning for LEO constellation networks [5–9]. Xiao et al. [5] proposed a LEO satellite network capacity model focusing on the influence of topology and routing strategy on throughput capacity. Qonita et al. [6] made the designation and analysis on multibeam satellite capacity for Indonesia. Arifn et al. [7] analyzed the achievable capacity of the Telesat system to provide service for Indonesian. del Portillo et al. [8, 9] compared the Telesat, OneWeb and Starlink networks and proposed a preliminary GS deployment solution based on non-dominated sorting genetic algorithm II (NSGA-II). Although the proposed NSGA-II based ground station planning scheme performs well, the NSGA-II method has quite high computational complexity. What's more, no interference is assumed, where the interference between multiple feeder links of same GS should not be ignored.

To solve this problem, this paper analyzes the influence of GS deployment on system throughput with interference considered. Meanwhile, a simple but effective iterative ground station deployment based on marginal revenue maximization (IGSD-MRM) algorithm is proposed to optimize ground segment. The main contributions of this paper are summarized as follows: (1) This paper considers the mutual interference between feeder links as the non-ideal characteristics of the antenna radiation will affect the capacity of the feeder link. (2) This paper no longer uses the metaheuristics algorithm (such as GA or NSGA-II) in GS planning, but designs an iterative method by exploiting the relationship between throughput and GS locations.

The rest of the paper is organized as follows: Section 2 describes the system model for LEO constellation scenario and formulates the optimization problem with the goal of maximizing system throughput. Section 3 presents the proposed IGSD-MRM algorithm. Section 4 describes the simulation parameters and results analysis. Section 5 concludes the paper.

2 System Model and Problem Formulation

2.1 System Model

For the LEO satellite constellation network, denote the number of LEO satellites, the number of orbital planes and the number of LEO satellites per plane as M, N_o, N_p, respectively. For each satellite, the position can be represented by the latitude and longitude of its sub-satellite point (SSP) and the orbital height. The SSP is defined as the intersection of the line linked the satellite and the center of the earth and the earth's surface. Considering that the existing LEO satellites are mostly circular orbits, the position of the satellite nodes is determined only by the latitude and longitude of the SSP. Since the satellite orbits the earth around the orbit and the earth is also rotating, the trajectory of the SSP will change with time. The longitude λ and latitude ϕ of the SSP is calculated through Eqs. (1) and (2), respectively:

$$\lambda = \lambda_p + \arctan(\cos i \tan u) - w_e t \qquad (1)$$

$$\phi = \arcsin(\sin i \sin u) \tag{2}$$

Where λ_p is the ascending longitude of the LEO satellite, i is the orbital inclination, w_e is the earth rotation angular velocity. Parameter u is the orbital angle during time t, $u = w_s \times t$, where the w_s is the satellite orbital moving angular velocity.

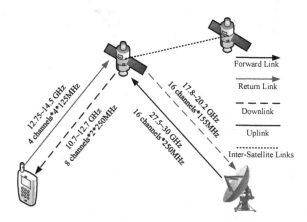

Fig. 1. The illustration of different links in LEO satellite networks

For each satellite, multiple beams are often generated in the coverage to achieve the spatial division gain, which has been proved great effectiveness in performance improvement [10–12]. The number of beams for each satellite is denoted as N. The links are mainly of the following three types: (1) User link (UL): the link between the satellite and the UTs; (2) Feeder link (FL): the link between the satellite and the GSs; (3) Inter-satellite link (ISL): the link between satellites. Meanwhile, the link can be divided into forward link and return link depending on the direction of service. The forward link refers to the link from the GSs to the UTs while the return link refers to the link from the UTs to the GSs. Figure 1 illustrates the aforementioned links taking OneWeb as example:

The achievable capacity of links is determined by two parameters, the link bandwidth and the spectrum efficiency (SE). The SE is mainly determined by link quality which further influences modulation and coding scheme (MCS). As an efficient transmission scheme in satellite communication systems, the Digital Video Broadcasting through Satellite Second Generation Extended (DVB-S2X) [13] adopts adaptive MCS and high-order modulation to cope with time-varying channel conditions. Figure 2 shows the SE of the DVB-S2X transmission scheme under different signal to interference plus noise ratio (SINR) with filter roll-off factor $\alpha = 0.1$.

From Fig. 2, we know that the SE varies with the receiving SINR. For LEO constellation network, the SINR can be calculated by Eq. (3):

$$SINR = EIRP - PL + G/T - I \tag{3}$$

Where the *EIRP* is the effective isotropic radiated power (EIRP) of the transmitting signal, *PL* is the propagation loss between the transmitter and the receiver (mainly

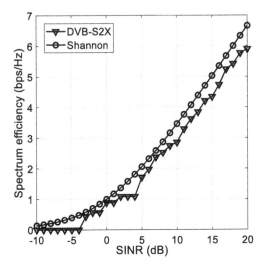

Fig. 2. The relationship between SE and SINR for DVB-S2X transmission scheme

includes free space propagation loss, rain attenuation and atmosphere absorption, etc.) [14], G/T is the quality of the receiving antenna. It should be noted here that the parameters in Eq. (3) are in form of dB.

Thus, the link achievable capacity can be calculated as Eq. (4):

$$C = B_c \times SE_{dvb-s2x}(SNR) \tag{4}$$

Where B_c is the link bandwidth, $SE_{dvb-s2x}(\cdot)$ is the spectrum efficiency mapping the SNR to SE under the DVB-S2X transmission scheme.

2.2 Problem Formulation

This paper aims at planning the ground station deployment so that the total system throughput is maximized. Without loss of generality, this paper considers the forward link. The system throughput is calculated and then the optimization problem is formulated.

For the LEO satellite constellation networks, the system throughput is the sum of capacity of each LEO satellite. In this paper, the bent-pipe mode is supposed for each LEO satellite, which is the same as OneWeb. The achievable capacity of LEO satellite is determined by the traffic demands, the capacity of feeder link and user link. Following is the method to calculate the aforementioned three factors:

(1) User link capacity

The user link capacity for satellite is the sum of all user links to multibeams, where user link capacity for each beam can be calculated through Eq. (4). Denote the user link capacity for beam b_n of satellite s_m as C_{s_m,b_n}^{UL}.

(2) Feed link capacity

The feeder link capacity can be calculated as Eq. (4), where the EIRP refers to the GSs transmitting EIRP, the G/T refers to the feeder receiving antenna of LEO satellite. Denote the feeder link capacity for beam b_n of satellite s_m as C_{s_m,b_n}^{FL}. It should be noted here that the interference I is mainly caused by the adjacent feeder links.

(3) Traffic demands

Traffic demands for beam b_n of satellite s_m mean the sellable capacity of LEO network, which is denoted as D_{s_m,b_n}.

Then, the total system throughput for LEO satellites constellation network is calculated as Eq. (5):

$$C_{tot} = \sum_{m=1}^{M} \sum_{n=1}^{N} \min\left(C_{s_m,b_n}^{UL}, C_{s_m,b_n}^{FL}, D_{s_m,b_n}\right) \tag{5}$$

From Eq. (5), we know that the system throughput depends on the link capacity and the traffic demands. Different ground station deployment mainly influences the achievable capacity of feeder link, which in turn influence the system throughput. Considering that the location of the ground station deployment is closely related to the link capacity, how to maximize the total constellation capacity under the constraints of the number of ground stations is of great significance and is also the goal of this paper.

The set of ground stations can be recorded as $\mathcal{G} = \{G_j | j = 1, 2, \cdots J\}$, where each ground station is identified by a pair of latitude and longitude, denoted as $G_j = [\phi_j, \ \theta_j]$. The number of satellites that can be connected simultaneously for each ground station is recorded as K. To this end, the optimization problem of the ground station deployment of LEO satellite constellation network is formulated as Eq. (6):

$$opt. \quad \mathcal{A} = \max_{\mathcal{G}} C_{tot} \tag{6}$$

For the optimization problem, the objective is to maximize the system throughput while the decision variables are the GS deployment. Considering that the location of the GS deployment is closely related to the link capacity, which is a discrete function of the link quality, to find an analytical method to achieve the optimal GS deployment is extremely challenging. Based on the aforementioned description, the next section proposes a heuristic method, namely the iterative GS deployment and marginal revenue maximization (IGSD-MRM) algorithm.

3 Proposed IGSD-MRM Algorithm

The main idea of the ground station optimization algorithm based on iterative greedy is to find the position of a single station at each iterative, and then do the next iteration based on the deployed situation of the station.

To optimize the GS deployment, the GS geographical location should consider the traffic demands and the link quality. The challenge lies in the discrete and non-linear of the formulated optimization problem. In this paper, the decision GS locations are limited to a set of discrete values by dividing the whole world into small grids. Then each grid represents a possible candidate for GS deployment and the grid which can achieve maximal marginal revenue is selected. Following is the process to decide the GS location:

(1) Divide the world into multiple grids where each grid X_i is 1° latitude by 1° longitude and granularity. Thus, total 360 * 180 = 64,800,000 grids are the candidates for GS deployment. All grid center point coordinates constitute a set as Eq. (7)

$$\mathcal{X} = \{X_i | X_i = [\phi_x, \theta_x]\} \tag{7}$$

(2) Calculate the traffic demands of the LEO satellite. For LEO satellite, its traffic demands are time-varying for its SSP are changing. The traffic demands for each grid can be denoted as $\mathcal{F}$ while the traffic demands for each satellite with its SSP X_i can be denoted as $\mathcal{F}_s(X_i)$.

(3) Calculate the marginal revenue for each possible GS deployment. For GS deployment $G_j = [\phi_j, \theta_j]$, calculate the region $\Phi(G_j)$ where LEO satellites can communicate with this GS satisfying the minimum FL elevation constraint. Marginal revenue is defined as the incremental throughput due to GS deployment and the marginal revenue is expressed as Eq. (8)

$$W(G_j) = \sum_{X_i \in \Phi(G_j)} w(X_i) \tag{8}$$

Following is the process to determine the parameter $w(X_i)$:

(a) Calculate the spherical distance between G_j and X_i;
(b) Calculate the corresponding geocentric angle for the spherical distance in (a);
(c) Calculating the propagation distance of the FL;
(d) Calculate the path loss PL based on the propagation distance and location-aware loss (such as rain attenuation);
(e) Calculate the link budget and SE as Eqs. (3) and (4);
(f) Calculate the achievable link capacity $C(X_i)$.
(g) Calculate the gridded revenue as Eq. (9):

$$w(X_i) = \min[C(X_i), \mathcal{F}_s(X_i)] \tag{9}$$

(4) Choose the location with maximal marginal revenue as the GS deployment. The GS location is expressed as Eq. (10)

$$X_i = \arg\max_{X_i} W(G_j) \tag{10}$$

4   Simulation Result and Analysis

In this section, the simulation result and analysis are given. Firstly, the simulation scenario and parameters are given. Then, the simulation process is illustrated. Finally, the simulation result and analysis are described.

4.1 Simulation Parameters

We mainly take OneWeb as the simulation scenario, where the LEO constellation consists of 720 satellites operating at 1200 km altitude. The inclination of 18 orbital planes is assumed 87°. The feeder uplink and the user downlink adopt the Ka and Ku bands, respectively [15]. The DVB-S2X scheme is supposed in both FL and UL. Main simulation parameters are listed in Table 1:

Table 1. Simulation parameters

Parameters	Values
LEO constellation	OneWeb
Number of satellites	720
Number of orbital planes	18
Number of satellites per plane	40
Orbital height	1200 km
Orbital inclination	87°
Minimum elevation for UTs	55°
Minimum elevation for GSs	20°
Uplink carrier frequency for FL	28.5 GHz
Uplink carrier bandwidth for FL	250 MHz
EIRP for transmitting GSs FL	52.0 dBW
G/T for receiving satellite FL	11.3 dB/K
Downlink carrier frequency for UL	13.5 GHz
Downlink carrier bandwidth for UL	250 MHz
EIRP for transmitting satellite UL	34.6 dBW
G/T for receiving UTsUL	12.9 dB/K
Number of GSs	[0, 30]
Number of feeder antennas per GS	5
GSs candidate location	Table 8 in [9]
Transmission mode for FL/UL	DVB-S2X
Roll off factor for DVB-S2X	0.1
Total traffic demands	400 Gbps
Traffic demands distribution model	Hotspot
Simulation steps	10 s
Simulation time	900 s
Grid point for LatLon	1°

The traffic model is assumed the hotspot model which is widely adopted for traffic demands in satellite networks [16]. Specifically, several hotspots are randomly generated in some parts of the world. The traffic demands is denoted as $\mathcal{F}$, indicating the density of traffic demands, in units of bps/m^2. The traffic demands is denoted as Eq. (11):

$$\mathcal{F} = \left\{ f(\phi, \theta) | \phi \in [-90°, 90°], \theta \in [-180°, 180°] \right\} \tag{11}$$

The traffic demand of the hotspot area is subject to the two-dimensional Gaussian model, and the hotspot area traffic demand generated in latitude ϕ and longitude θ is calculated as Eq. (12):

$$f(\phi, \theta) = \frac{1}{2\pi \sigma_\phi \sigma_\theta \sqrt{1 - \rho_{\phi,\theta}^2}} \cdot$$
$$e^{\left[-\frac{1}{2(1-\rho_{\phi,\theta}^2)} \left\{ \frac{(\phi-\phi_0)^2}{\sigma_\phi^2} + \frac{(\theta-\theta_0)^2}{\sigma_\theta^2} - \frac{2\rho_{\phi,\theta}(\phi-\phi_0)(\theta-\theta_0)}{\sigma_\phi \sigma_\theta} \right\} \right]} \tag{12}$$

Where the (ϕ_0, θ_0) is the latitude and longitude of the hot spot.

Here, total 1860 hot spots are selected according to the maximum population cities in the late 2018 statistics. What's more the traffic demand is proportional to its population. The traffic demands $\mathcal{F}$ used in the simulation is shown as Fig. 3.

Fig. 3. Traffic demands throughout the world

The service requirements in each beam coverage area for each LEO satellite can be calculated by Eq. (13):

$$D_{s_m,b_n} = \sum_{(\phi,\theta) \in \Phi_{s_m,b_n}} f(\phi, \theta) \cdot d_{\phi,\theta} \tag{13}$$

Where $d_{\phi,\theta}$ is the difference area at geographical location (ϕ, θ), Φ_{s_m,b_n} is the set of geographical locations in satellite s_m beam b_n coverage.

4.2 Simulation Process

The system throughput of LEO satellite constellation networks is evaluated through the "Low Earth Orbit Satellite Constellation Network Planning and Network Optimization

Platform" (LEO-NPNOP). The LEO-NPNOP focuses on the next generation of mega LEO satellite constellations such as OneWeb. The LEO-NPNOP can implement the network planning in the pre-stage and network optimization in post-stage, considering different constellation configurations, ground segment deployment, satellite/beam resource scheduling and FL/UL/ISLs bandwidth and link budget, etc. The Monte Carlo method is adopted. The process of the LEO-NPNOP is illustrated as Fig. 4:

4.3 Simulation Results

(1) Performance under different number of GSs

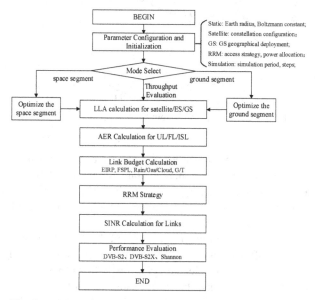

Fig. 4. System throughput evaluation process in LEO-NPNOP

Figure 5 shows the system throughput varies with the number of GSs. The number of FL antennas is fixed 10.

It can be seen from Fig. 5 that the system throughput of LEO satellite constellation networks increases with the number of ground stations. The proposed IGSD-MRM algorithm is significantly better than hotspot based GS deployment scheme. Take number of GS 20 as example, the system throughput of the proposed IGSD-MRM scheme and the hotspot-based scheme can achieve 299.9 Gbps and 141.7 Gbps, respectively. That is to say, the proposed scheme can increase the system throughput by 111.6%. Meanwhile, the growing rate of system throughput has slowed down with the increase in the number of GSs. This can be explained by the fact that more GSs are deployed in non-hotspot areas in large number of GSs.

Figure 6 shows the incremental throughput with the increase of the number of GSs. It can be seen from Fig. 6 that the incremental throughput decreases as the number of GSs increases. This is because when the number of GSs is small, the system throughput can be heavily increased by deploying more GSs in the hotspot area.

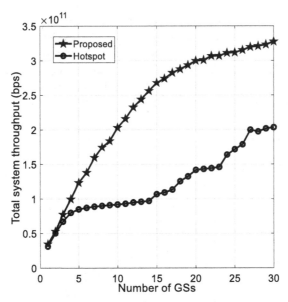

Fig. 5. Relationship between system throughput and number of GSs

While as the number of GSs increases, more GSs are in charge of areas with small traffic demands. For example, the incremental system throughput is about 18 Gbps when the number of GSs is 1, that is to say, the single GS can provide 18 Gbps system throughput. However, the single GS can provide only 3 Gbps when the number of GSs is more than 22.

Based on the traffic demand model in Fig. 3 and the described method in Sect. 3, the LEO satellite traffic demands can be calculated and showed in Fig. 7. Figure 8 shows the candidate GSs and selected GS through the proposed algorithm.

(2) Performance under different number of FL antennas per GS

Figure 9 shows the system throughput varies with the number of FL antennas per GS. The number of GSs is fixed 30.

It can be seen from Fig. 9, the system throughput increases as the number of FL antennas increases while the growing rate is slower as the number of FL antennas increases. When the number of FL antennas increases from 2 to 4, the system throughput increases from 102.8 Gbps to 190.4 Gbps. That is to say, the throughput increases by 85.21% when the number of feeder antennas increases by 100%. However, when the number of FL antennas increases from 8 to 16, the system throughput increases from 297.5 Gbps to 328.9 Gbps. That is to say, there is only 10.6% improvement even the number of FL antennas is doubled.

From Fig. 9, the system throughput keeps nearly steady when the FL antennas per GS exceed 14. This can be explained as follows: For the OneWeb constellation networks, there is certain number of LEO satellites can be accessed for each GS under 20° elevation constraint. No more LEO satellite can be linked for given GSs even if there is more FL antennas per GS.

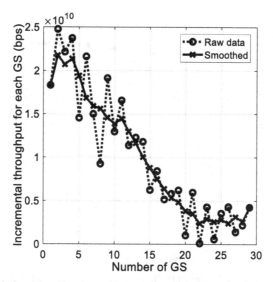

Fig. 6. Relationship of incremental throughput for each GS and number of GSs

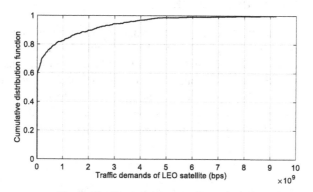

Fig. 7. Traffic demands for LEO satellites

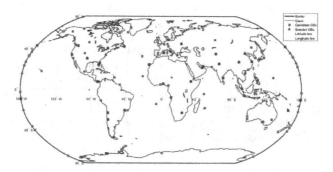

Fig. 8. The GSs deployment through IGSD-MRM algorithm

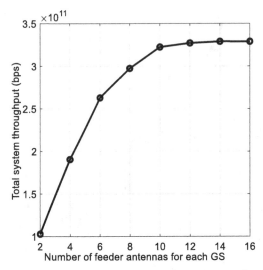

Fig. 9. Relationship of incremental throughput for each GS and number of GSs

5 Conclusion and Future Work

This paper describes the system throughput evaluation and proposes a heuristic method to optimize ground station deployment for the LEO constellation networks. Considering the unbalanced traffic demands around the world, the location of the GSs deployment is critical to improve the system throughput. The number of GSs and the number of feeder link antennas per GS should be rationally designed to maximize the system throughput. Our next work is to evaluate the system throughput and optimize ground station deployment for LEO constellation network with ISLs.

References

1. Pultarova, T.: Telecommunications-space tycoons go head to head over mega satellite network. News Brief. Eng. Technol. **10**(2), 20 (2015)
2. Liu, S., Hu, X., Wang, Y., Cui, G., Wang, W.: Distributed caching based on matching game in LEO satellite constellation networks. IEEE Commun. Lett. **22**(2), 300–303 (2018)
3. Christensen, C., Beard, S.: Iridium: failures & successes. Acta Astronaut. **48**(5–12), 817–825 (2001)
4. Su, Y., Liu, Y., Zhou, Y., Yuan, J., Cao, H., Shi, J.: Broadband LEO satellite communications: architectures and key technologies. IEEE Wirel. Commun. **26**(2), 55–61 (2019)
5. Xiao, Y., Zhang, T., Shi, D., Liu, F.: A LEO satellite network capacity model for topology and routing algorithm analysis. In: 2018 14th International Wireless Communications & Mobile Computing Conference (IWCMC), Limassol, pp. 1431–1436 (2018)
6. Qonita, A.H., Muhtadin, N.: Design and analysis of multibeam communication satellite links operated at Ka Band Frequency in Indonesia. IOP Conf. Ser. Earth Environ. Sci. **284**(1), 012049 (2019)
7. Arifn, M.A., Khamsah, N.M.N.: A case study in user capacity planning for low earth orbit communication satellite. In: 2018 IEEE International Conference on Aerospace Electronics and Remote Sensing Technology (ICARES), Bali, pp. 1–6 (2018)

8. del Portillo, I., Cameron, B.G., Crawley, E.F.: A technical comparison of three low earth orbit satellite constellation systems to provide global broadband. Acta Astronaut. **159**, 123–135 (2019)

9. del Portillo, I., Cameron, B., Crawley, E.: Ground segment architectures for large LEO constellations with feeder links in EHF-bands. In: 2018 IEEE Aerospace Conference, Big Sky, pp. 1–14 (2018)

10. Liu, S., Hu, X., Wang, W.: Deep reinforcement learning based dynamic channel allocation algorithm in multibeam satellite systems. IEEE Access **6**, 15733–15742 (2018)

11. Hu, X., et al.: Deep reinforcement learning-based beam Hopping algorithm in multibeam satellite systems. IET Commun. **13**(16), 2485–2491 (2019)

12. Hu, X., Liu, S., Chen, R., Wang, W., Wang, C.: A deep reinforcement learning-based framework for dynamic resource allocation in multibeam satellite systems. IEEE Commun. Lett. **22**(8), 1612–1615 (2018)

13. Digital Video Broadcasting (DVB): Second generation framing structure, channel coding and modulation systems for broadcasting; Part2: DVB-SE extensions (DBVS2X), Technical report 6 (2014)

14. ITU: Propagation data and prediction methods required for the design of earth-space telecommunication systems, Recommendation ITU-R 618–13 (2017)

15. WorldVu Satellites Limited: OneWeb Ka-band NGSO constellation FCC filing SAT-LOI-20160428-00041, 12 September 2012. http://licensing.fcc.gov/myibfs/forwardtopublictabaction.do?file_number=SATLOI2016042800041

16. Nishiyama, H., Tada, Y., Kato, N., Yoshimura, N., Toyoshima, M., Kadowaki, N.: Toward optimized traffic distribution for efficient network capacity utilization in two-layered satellite networks. IEEE Trans. Veh. Technol. **62**(3), 1303–1313 (2013)

Deep Learning Based Intelligent Congestion Control for Space Network

Kun Li, Huachun Zhou$^{(\boxtimes)}$, Hongke Zhang, Zhe Tu, and Guanglei Li

School of Electronic and Information Engineering,
Beijing Jiaotong University, Beijing 100044, China
`{19111021,hchzhou,hkzhang,zhe_tu,15111035}@bjtu.edu.cn`

Abstract. In order to alleviate the impact of network congestion on the spatial network running traditional contact graph routing (CGR) algorithm and DTN protocol, we propose a flow intelligent control method based on deep convolutional neural network (CNN). The method includes two stages of offline learning and online prediction to intelligently predict the traffic congestion trend of the spatial network. A CGR update mechanism is also proposed to intelligently update the CGR to select a better contact path and achieve a higher congestion avoidance rate. The proposed method is evaluated in the prototype system. The experimental results show that it is superior to the existing CGR algorithm in terms of transmission delay, receiver throughput and packet loss probability.

Keywords: Contact graph routing · Space network · Deep convolutional neural network · Intelligent congestion control

1 Introduction

Recently, due to the increasing number of satellite nodes and the increasing complexity of satellite applications, the traffic has increased dramatically in the space network. The rapid growth of network traffic and the increasing variability are bringing great pressure to the space network, thus affecting the Quality of Experience (QoE) of users. Although the Delay Tolerant Network (DTN) [1], which is widely used in satellite networks, is designed to cope with frequent link interruptions, long end-to-end delays, and high channel error rates in the network. However, when the network suffers from excessive traffic load, the satellite nodes running the DTN protocol will be discarded due to insufficient storage space, which will cause network congestion. The traditional method is to deploy the Contact Graph Routing (CGR) in the satellite network to take advantage of the predictable characteristics of the spatial node trajectory, and to use the known connection time and the remaining storage space of the predetermined neighboring nodes to determine the effective route, but the CGR does not consider the remaining storage space of other nodes in the path. When the remaining storage space of the intermediate node is lower than the contact remaining capacity, traffic congestion will still occur [2]. In addition, due to the periodic nature of the spatial network, when severe congestion occurs again, the traditional routing strategy will not learn and improve from previous

© Springer Nature Singapore Pte Ltd. 2020
Q. Yu (Ed.): SINC 2019, CCIS 1169, pp. 16–27, 2020.
https://doi.org/10.1007/978-981-15-3442-3_2

problems such as high latency and high packet loss. Therefore, it is necessary to learn spatial network congestion scenarios in an intelligent way in order to manage large-scale growth of network traffic better in the space network.

We propose a space network intelligent congestion control method based on deep learning. The offline learning phase uses the spatial information dataset extracted and constructed from the prototype system as the input and output features of the Convolutional Neural Network (CNN). The neural network structure is used to train adaptive network traffic characteristics. In the online prediction phase, combined with the idea of Software Defined Satellite Networks (SDSN) [3], the trained CNN model is deployed in the GEO satellite as an intelligent control node, and the packet rate sent from the MEO satellite node is collected in real time as input data. Through the CNN model to predict the trend of network traffic changes, a CGR update mechanism is proposed to intelligently change the CGR to select a better contact path, thereby alleviating the traffic congestion problem in the space network. The performance of the proposed method is evaluated in the prototype system, which proves that it is superior to the existing CGR algorithm in terms of transmission delay, packet loss rate and receiver's throughput.

The rest of the paper is organized as follows: In Sect. 2, we discuss the related work. In Sect. 3, we present CNN-based intelligent congestion control methods and prototype implementations, and then performs performance analysis in Sect. 4. At last, Sect. 5 summarizes the paper and future work.

2 Related Work

Recently, many studies have focused on the application of machine learning in the field of network flow control systems or DTN route optimization. Regarding the former, Mao et al. [4] propose an intelligent packet routing strategy based on Tensor's Deep Belief Architecture (TDBA), which takes into account multiple parameters of network traffic to predict the full path of each edge router in the wireless backbone network. Hendriks et al. [5] proposed a Q-Routing algorithm, which combines the existing technology in wireless routing and optimizes the routing method by using multi-agent reinforcement learning technology. However, due to the high mobility of satellite nodes and frequent link switching, these methods cannot be applied to space networks. We consider the characteristics of space network and combines it with deep learning to control the traffic between satellite nodes intelligently. Different from the above work, the complex characteristics of the aerospace integrated network are analyzed [6], and a deep learning-based method is proposed to improve the satellite flow control performance. However, the simulation experiment did not combine the characteristics of the satellite network, it is difficult to verify the effect of the method, and we work in the DTN-based prototype system to verify the effectiveness of the method.

In addition, several efforts apply machine learning techniques to routing optimization in DTN [7, 8]. Dudukovich et al. [7] adopted the methods of Q routing and naive Bayes classification to develop the architecture of the cross-layer feedback framework of DTN protocol, so as to select the transmission path with lower delay. Papachristou et al. [8] proposed to use a machine learning classifier to predict a set of neighboring nodes, which are most likely to pass messages to the desired location based on the message history to determine the transmission path of the packet. However, these DTN-based route optimization methods focus on low-latency path selection, but the work of this paper focuses on traffic congestion control in space networks.

We propose a space network intelligent congestion control method based on deep learning. It aims to use the feature extraction ability of deep learning to extract the traffic information of satellite nodes and predict the traffic congestion in real time, and then dynamically adjust the CGR algorithm to avoid traffic congestion to reduce the data transmission delay in the space network.

3 Traffic Intelligent Control Method Based on CNN and Prototype Implementation

We combine the idea of SDSN to study the intelligent congestion control method in space network. The following is an analysis of the intelligent space network management architecture design and prototype system implementation.

3.1 Intelligent Space Network Management Structure

In order to briefly describe the traffic congestion problem of the space network, we propose a two-layer satellite network composed of two MEO satellites and two LEO satellites, as shown in Fig. 1. The data is transmitted from two MEO satellites to the LEO2 satellite at the same time. Before the MEO satellite bursts, the load of the MEO2 satellite is very small. Therefore, the traditional CGR algorithm running in the satellite network selects the data sent by the MEO satellite to be forwarded through the MEO2 satellite node. However, when the sudden overflow occurs in the source MEO1 node, the MEO2 satellite load increases sharply and serious congestion occurs. At this time, the CGR algorithm first transmits the packet with higher priority, but as the packet loss rate increases, the alternative path is selected (such as forwarding through the LEO1 satellite) to alleviate the congestion of the MEO2 satellite node, but the data loss situation It is very serious. At the same time, when the same congestion occurs again in the space network as the cycle runs, the CGR algorithm still uses the same processing method to combat similar congestion problems, and the space network congestion condition cannot be solved. Therefore, we propose an intelligent congestion control method in space network based on the idea of SDSN. The core idea is to deploy a deep learning method with feature extraction capability in the spatial network to process the spatial data set composed of the traffic information collected during the operation cycle and the corresponding network congestion. The network congestion problem is avoided by fully extracting the identifiable features in the traffic information to predict the occurrence of network congestion in a similar situation and intelligently switching the transmission path with reference to the CGR update mechanism. Since the traffic information in the spatial network can be characterized as a two-dimensional matrix and the number of samples is large, we select the deep CNN which is good at processing high-dimensional data and large sample size as the traffic prediction tool deployed in the intelligent controller.

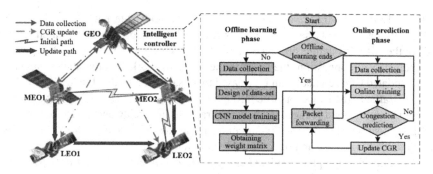

Fig. 1. Intelligent space network architecture

The intelligent controller is deployed in the Geosynchronous Earth Orbit (GEO) satellite, and the Middle Earth Orbit (MEO)/Low Earth Orbit (LEO) satellite is responsible for data forwarding in the space network. The satellite traffic is sent by the MEO satellite to the LEO2 satellite, and the transmission path matches the CGR. At the same time, GEO satellites establish interfaces with MEO satellites to monitor MEO satellite transmission traffic and collect them in real time. It is processed into an input feature matrix recognizable by the CNN model, trained by the CNN model and predicting network congestion. If it is determined that congestion occurs, the CGR contact plan is adjusted with reference to the neural network output information, thereby changing one or more transmission paths to avoid network congestion. Figure 1 shows the congestion control process of the intelligent space network based on the SDSN idea.

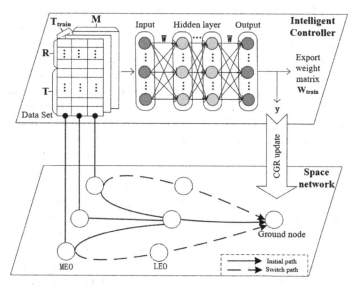

Fig. 2. Offline training and online prediction based on CNN

The MEO1 and MEO2 satellites simultaneously send data packets to the LEO2 satellite. When the GEO satellite collects the traffic sent by the MEO satellite and predicts the MEO2 satellite presence network through the CNN model. After getting the conclusion of congestion, the CGR in the space network is automatically updated and the updated CGR contact plan is released to all satellites, so that the MEO1 satellite transmits traffic to the LEO2 satellite through the LEO1 satellite to avoid congestion on the MEO2 satellite, and reduces the network transmission delay and packet loss rate.

Figure 2 shows that the intelligent controller in the GEO satellite has two phases: the offline learning phase and the online prediction phase.

The offline learning phase: When deploying an intelligent controller in a space network, it is first in the offline learning phase, which uses past traffic patterns to train the CNN model to predict the traffic load in the future space network. In the initial state, the traditional CGR algorithm is used for path selection in the MEO/LEO satellites. The GEO satellite collects the packet rate v_i from each MEO satellite in real time through the interception interface and continuously collects the data of the duration $T_{continued}$ as the original data for the CNN training. In order to make better use of the feature extraction ability of deep CNN, we describe the input data format as a two-dimensional matrix (M, T), where $M = \{1, 2, \cdots, i\}$ represents different MEO satellites, $T = \{1, 2, \cdots, T_{continued}\}$ represents the duration of a data collection, and the value of column t of the i-th row represents the packet rate v_{it} of the i-th MEO satellite at the t s. Therefore, each row is used to record the flow characteristics of different MEO satellites, and each column is used to record flow characteristics at different points in time. Since the spatial network has a long delay and a high channel error rate, $T_{continued}$ is set to 15 s to ensure that the CNN model has enough input information to extract the traffic characteristics. The traffic transmission paths of all MEO satellites are stored as a two-dimensional matrix (M, R), where $R = \{s_{i1}, s_{i2}, \cdots, s_{iN}\}$ represents the routing path of the source i-th MEO satellite, $N = \{1, 2, \cdots, i, \cdots, j\}$ represent different MEO/LEO satellites. The lengths of different MEO satellite path are different. Select the longest transmission path R_{max} in the network topology as the standard form of the matrix (M, R_{max}), and the remaining paths use 0 to complete the value of the matrix. Finally, the link delay is detected to determine whether the network is congested during the $T_{continued}$ period. The result is recorded as $\{y|0, 1\}$. It is detected that the congestion occurs immediately after the threshold is exceeded and the record is 1; otherwise, it is recorded as 0. The traffic characteristics and congestion states are eventually combined into a labeled training set $D = \{x(M \times (T + R)), y\}$. The duration of the entire offline learning phase is T_{train}, and the spatial dataset is stored as a three-dimensional tensor $(M, T + R, T_{train})$. In order to use sufficient time slots to ensure the completeness of the data set, T_{train} is set to 2 days. Since the space data set records the packet information and the routing path of the space network source node in the past time T_{train} and combines them into a three-dimensional tensor format, therefore, the data set can be regarded as a two-dimensional feature grayscale image of size $(M, T + R)$ in each time slot of T_{train}. Since the CNN is identified based on the texture features and the satellite operation is periodic, the traffic gray level map composed of the traffic at the time of the congestion and the path can be identified by the CNN and associated with the congestion condition represented by the label y. After multiple iterations, The weight matrix W of the hidden layer in the CNN model is adjusted to W_{train} by forward propagation and gradient descent. When the next cycle arrives, the characteristic grayscale image based on traffic and path can achieve high prediction accuracy under this weight matrix, and W_{train} is derived as the final model of the online prediction phase.

The online prediction phase: the trained CNN weight matrix W_{train} is imported into the intelligent controller in the GEO satellite, and the traffic characteristics of each MEO satellite are collected in real time through the interception interface and the input information is characterized as a two-dimensional matrix $(M, T + R)$. The data set format is adapted to the CNN weight matrix W_{train} for online prediction. At the same time, the alternate paths of all MEO satellites are stored as a two-dimensional matrix (M, R_{backup}) to cope with the update mechanism after congestion prediction. According to the predictive output y of the CNN model, we propose a CGR update mechanism based on CNN congestion prediction, which adopts a path-by-path update method to dynamically adjust the CGR to avoid serious packet loss caused by frequent path switching. The specific implementation is described in the algorithm shown in Table 1.

Table 1. CGR update mechanism based on CNN congestion prediction

1 The controller interface collects the traffic characteristics and inputs the trained CNN weight matrix W_{train}
2 If CNN model predicts output y to be 0
3   Controller interface continues to monitor incoming traffic information
4 Else
5   While CNN model predicts output y to be 1
6   Obtain the packet rate $v_{i\,T_{continued}}$ at the $T_{continued}$ time in the traffic matrix $(M, T + R)$ and sort it to obtain the MEO satellite with the highest packet rate in the i MEO satellites, and update the CGR of the space network with reference to the path matrix (M, R_{backup}). To match the alternate path of the corresponding MEO satellite
7   Remove the MEO satellite from the i MEO satellite queue and update the path matrix to $(M - 1, R_{backup})$
8   The controller interface continues to collect traffic for duration $T_{continued}$
9   Controller interface continues to listen to traffic information

3.2 Prototype Implementation

We use the standard Tr constellation [9], which consists of three GEO satellites (GEO1–GEO3), 10 MEO satellites (MEO1–MEO10) and 66 LEO layer satellites (LEO1–LEO66). Table 2 shows the specific parameters of the Tr constellation.

In addition, we also set up a ground station——Beijing Station (116°E, 40°N), which is used to measure the link performance between the satellite network and the ground network under high-speed mobile state. A GEO satellite is continuously connected to the Beijing Railway Station. The MEO satellite not only maintains continuous connection with the GEO layer satellite, but also maintains continuous connection with the LEO layer satellite. At the same time, some LEO satellites continuously cover the Beijing Station.

Table 2. Three-layer satellite network parameters

	Track height/km	Operating cycle/h	Number of satellites	Orbital inclination/°
GEO satellite	36000	24	1×3	0
MEO satellite	10390	6	2×5	45
LEO satellite	895.5	12/7	6×11	90

The prototype uses the DTN protocol developed by NASA to implement software ION-3.5.0 [10] to meet the high transmission delay requirements, and the CGR has been implemented in it. The intelligent controller based on TensorFlow [11] is embedded in the GEO satellite node, and the shell script is written and installed on the MEO/LEO satellite node to generate traffic at the source node. The GEO satellite and MEO/LEO satellite realize data transmission through the Socket interface. CGR update.

4 Performance Analysis

This section will introduce the experimental design and analyze the experimental results.

4.1 Experimental Design

Two topologies were deployed in this lab to test the effectiveness of the traffic congestion control method in the space network. Topology 1 is shown in Fig. 3.

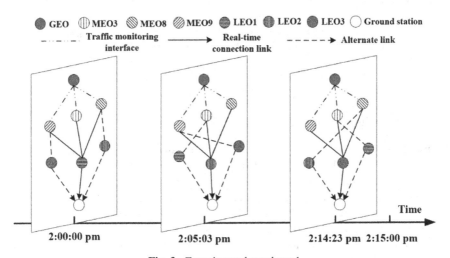

Fig. 3. Experimental topology 1

The connection between the Beijing station and the LEO satellite is constantly changing over time. Considering the high traffic rate after 2:00 pm, from 2:00:00 pm to 2:1 pm, there are 3 LEO satellites in the first LEO track gradually covering the Beijing station for 900 s. Based on the shortest link distance between LEO satellite and Beijing station, LEO1, LEO2 and LEO3 satellites are selected as relay satellites in different durations. The handover takes place at 2:05:03 (303 s) and 2:14:23. (863 s). MEO3, MEO8 and MEO9 satellites are selected as the source node for the distribution of 10 MEO satellites. Since these three MEO satellites can cover all transit LEO satellites from 2:00:00 to 2:15:00, GEO satellites can always cover all Satellite and ground stations. Figure 3 shows the flow of data between satellites during the experimental period of 0–900 s.

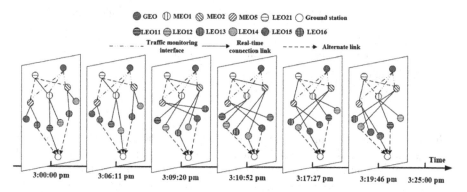

Fig. 4. Experimental topology 2

Topology 2 considers the problem of data transmission between LEO satellites across tracks, as shown in Fig. 4. From 3:00:00 pm to 3:25:00 pm, a total of 6 LEO satellites in the first LEO track gradually covered the Beijing station for a total of 1500 s. The LEO satellites are still set as relay satellites in different experimental time based on the shortest link distance from Beijing station. The switching occurs at 3:06:11 (371 s), 3:09:20 (560 s), 3:10:52 (652 s), 3:17:27 (104 s), and 3:19:46 (1186 s). Select a LEO satellite as the originating node from the second LEO orbit, MEO1, MEO2 and MEO5 satellites as relay satellites, and always maintain communication with the source LEO satellite nodes. While considering the limited number of satellite interfaces, each MEO satellite only It can be connected to the two LEO satellites in the first orbit, and the GEO satellite can always cover all satellites and ground stations.

In this paper, deep CNN is used as the deep learning structure deployed in GEO satellite intelligent controller. In the offline learning phase, 11520 traffic matrices are collected as spatial datasets, and the training phase is iterated 2000 times to fully extract the traffic characteristics in the space network. At the same time, in order to provide the appropriate CNN structure according to the spatial traffic matrix, we propose and compare five CNN architectures from shallow to deep, consisting of single, two, three, four and eight convolution and pooling layers. Tested by space network datasets, single-layer CNNs exhibit lower prediction accuracy due to poor feature extraction capabilities. While CNN models above three layers have long-term training accuracy, but the training time is too long. The accuracy in real-time judgment of spatial traffic decreases, which is caused by the over-fitting problem of deep networks. Therefore, the CNN structure used in this paper is a feature extraction component composed of two convolutional layers to filter the input features. After each convolutional layer, a pooling layer is set to cut the features into several regions, and the maximum value is taken to obtain the dimension. A smaller feature that combines all local features into global features through two fully connected layers to perform the classification process and give the final classification output, achieving a higher (90.36%) learning accuracy.

Table 3. Link delay parameter ·

	GEO satellite	MEO satellite	LEO satellite	Beijing station
GEO satellite	\	86 ms	116 ms	\
MEO satellite	86 ms	66 ms	50 ms	\
LEO satellite	116 ms	50 ms	\	3 ms\
Beijing station	\	\	3 ms	\

In order to make the simulation feasible and close to the real scene, we consider the actual delay of the satellite link by the parameters abstracted from STK [12], and then use the flow control tool of Linux to set the actual delay of the satellite link. In the experiment, the inter-satellite link rate is set to 250 kB/s, and the bit error rate is set to 10^{-6}. Table 3 shows the specific delay of each link in the topology.

4.2 Experimental Result

In this section, the performance of the spatial network intelligent congestion control method in Topology 1 and the traditional CGR algorithm are first compared. To simulate network congestion, random burst input traffic data is taken at each MEO satellite node, and the rate at which packets arrive at the LEO satellite interface is shown in Fig. 5(a). At 0–200 s, the arrival rate is always lower than the maximum capacity of the inter-satellite link, and there is basically no congestion. At 200–400 s, the packet transmission rate far exceeds the inter-satellite link rate. The satellite network will cause serious traffic congestion. The arrival rate recovers smoothly in the 400–600 s period, allowing the DTN protocol to retransmit data to alleviate the network performance degradation caused by traffic congestion. At 600–900 s, the arrival rate experiences two surges and falls back to distinguish the long-term traffic congestion.

Firstly, the data packets sent by the three MEO satellites are characterized as the traffic matrix as the input of the CNN model, and the obtained output predicted value y is shown in Fig. 5(b). When the predicted value y is 1, the congestion of the predicted space network is about to occur. At this time, the standby link is switched according to the CGR update mechanism, which occurs in the 165 s, 240 s, 331 s, 482 s, 537 s, and 680 s respectively. The satellite link has been switched a total of six times.

Then the experiment measured the throughput of the Beijing station node and the transmission delay from the MEO1 satellite to the Beijing station. The throughput and delay of the two methods are compared as shown in Fig. 5(c) and (d), and the experimental data was captured by the Wireshark tool installed on the Beijing station node.

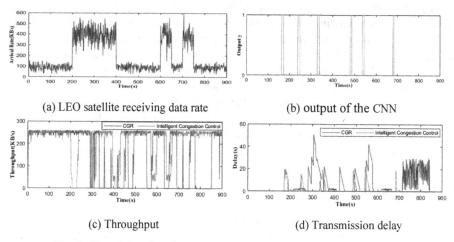

(a) LEO satellite receiving data rate   (b) output of the CNN

(c) Throughput   (d) Transmission delay

Fig. 5. Experimental performance comparison and testing in topology one

The experiment also tested the packet loss rate during the transmission of data from three MEO nodes to the Beijing station. The CGR algorithm has a total packet loss rate of 39.44%, which is caused by the lack of storage space of the relay LEO node due to network congestion. The space network using the intelligent congestion control method will fully utilize the storage space of the standby node. The packet rate is reduced to 17.33%, which improves the reliability of data transmitted by the space network.

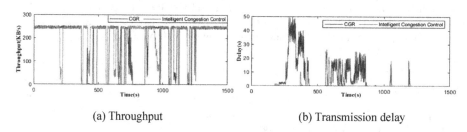

(a) Throughput   (b) Transmission delay

Fig. 6. Experimental performance comparison and testing in topology two

In the topology two, the rate in Fig. 5(a) is still used as the transmission rate of the source LEO satellite node. In the experimental time of 900–1500 s, the transmission rate is always stable and lower than the maximum capacity of the inter-satellite link. The throughput of the Beijing station node and the transmission delay from the source LEO satellite to the Beijing station are measured separately. The throughput and delay of the two methods are compared as shown in Fig. 6(a) and (b). In the normal communication situation of 0–200 s, the measured throughput of both methods is maintained at about 250 Kbps, and the delay is maintained within 200 ms. After entering the network congestion state after 200 s, the space network throughput using the CGR algorithm is significantly reduced. With the delay continuously increasing, the intelligent congestion control method can maintain the network with high throughput and low latency. Throughout the experiment, the intelligent congestion control method improved the throughput of the traditional CGR algorithm by 20.66% and the transmission delay by 84.61%, which improved the transmission performance of the space network. Due to frequent network connection switching, the space network using the CGR algorithm causes a packet loss rate of about 46.71%, and the space network using the intelligent congestion control method reduces the packet loss rate to 29.53%.

According to the above experiments, the intelligent congestion control method proposed is more effective than the traditional CGR algorithm, which alleviates the large amount of packet loss and high delay caused by congestion in the space network. It is an effective method for controlling traffic congestion in the space network.

5 Conclusion

Aiming at the difficulty of predicting and dealing with traffic congestion in traditional space networks, we propose a method based on deep CNN for intelligent congestion control and implement it in the prototype system. The experimental results show that the intelligent congestion control method can predict most of the congestion in multiple scenarios in a periodically operating spatial network, and dynamically adjust the CGR to improve the reliability of the data transmitted by the space network. The performance of the packet loss rate is better than the traditional CGR method, so as to alleviate the congestion of the space network. In the future work, we will focus on more efficient deep learning architectures and algorithms, and how to deploy intelligent congestion control methods in large-scale spatial network topologies.

Acknowledgement. This paper is supported by National Key R&D Program of China under Grant No. 2018YFA0701604, NSFC under Grant No. 61802014, No. U1530118, and National High Technology of China ("863 program") under Grant No. 2015AA015702.

References

1. Cerf, V., Burleigh, S., Hooke, A., et al.: Delay-tolerant networking architecture [S/OL] (2007). https://www.rfc-editor.org/info/rfc4838
2. Shi, W., Gao, D., Zhou, H.: QoS based congestion control for space delay/disruption tolerant networks. J. Electron. Inf. Technol. **38**, 2982–2986 (2016)
3. Li, T., Zhou, H., Luo, H., Yu, S., et al.: SERvICE: a software defined framework for integrated space-terrestrial satellite communication. IEEE Trans. Mob. Comput. **17**(6), 703–716 (2018)
4. Mao, B., et al.: A tensor based deep learning technique for intelligent packet routing. In: GLOBECOM 2017 – 2017 IEEE Global Communications Conference, pp. 1–6 (2017)
5. Hendriks, T., Camelo, M., Latré, S.: Q2-routing: a Qos-aware Q-routing algorithm for wireless ad hoc networks. In: 2018 14th International Conference on Wireless and Mobile Computing, Networking and Communications (WiMob), Limassol, pp. 108–115 (2018)
6. Kato, N., et al.: Optimizing space-air-ground integrated networks by artificial intelligence. IEEE Wirel. Commun. **26**, 140–147 (2019)

7. Dudukovich, R., Hylton, A.: A machine learning concept for DTN routing. In: 2017 IEEE International Conference on Wireless for Space and Extreme Environments (WiSEE), Montreal, pp. 110–115 (2017)
8. Dudukovich, R., Papachristou, C.: Delay tolerant network routing as a machine learning classification problem. In: 2018 NASA/ESA Conference on Adaptive Hardware and Systems (AHS), Edinburgh, pp. 96–103 (2018)
9. Long, F.: Satellite network robust QoS-aware routing. Springer, Heidelberg (2014). https://doi.org/10.1007/978-3-642-54353-1
10. Interplanetary Overlay Network (ION) [OL] (2018). https://sourceforge.net/projects/ion-dtn/
11. TensorFlow (2019). https://tensorflow.google.cn/
12. Satellite Tool Kit (STK) [EB/OL] (2018). https://www.agi.com/products/stk/

Multilayer Satellite Network Topology Design Technology Based on Incomplete IGSO/MEO Constellation

Liang Qiao[1,2(✉)], Hongcheng Yan[1,2], Yahang Zhang[1,2], Rui Zhang[1,2], and Weisong Jia[1,2]

[1] Institute of Spacecraft System Engineering, China Academy of Space Technology, Beijing 100094, China
owenqiao@126.com

[2] Science and Technology on Communication Networks Laboratory, Beijing 100094, China

Abstract. To meet the needs of step-by-step construction of inclined geosynchronous orbit (IGSO)/medium earth orbit (MEO) constellation in the future and to improve the robustness of the constellation in case of losing connections of some satellites, this paper studies the incomplete IGSO/MEO constellation network topology design technology, analyses the inter-satellite link accessibility, and proposes a multilayer satellite network link-building strategy based on dynamic programming. This paper also designs the methods of link-building for intra-layer links among IGSO satellites, intra-layer links among MEO satellites and inter-layer links between IGSO satellites and MEO satellites. The simulation results show that this method can well construct dynamic inter-satellite network topology with less link switches, higher network transmission bandwidth and wider access user coverage.

Keywords: Incomplete IGSO/MEO constellation · Network topology · Dynamic programming

1 Introduction

Because multi-layer satellite network can combine the advantages of both high-orbit satellites and medium-low-orbit satellites, it has become a research hotspot of space network. The construction of space network based on inclined geosynchronous orbit (IGSO)/medium earth orbit (MEO) constellation has higher low-cost advantages and better application prospects. In order to meet the needs of step-by-step construction of IGSO/MEO constellation in the future, and to improve the self-repairing ability of the constellation in case of disconnection of some satellites, and to enhance the reliability and robustness of the whole constellation, this paper mainly studies the network topology design of incomplete IGSO/MEO constellation. The satellites in incomplete IGSO/MEO constellation network cannot be kept visible to each other at any time. Inter-satellite links need to be switched dynamically, and the topology structure changes dynamically [1]. Therefore, it is necessary to study the link-building strategy for intra-layer links among

IGSO satellites, intra-layer links among MEO satellites and inter-layer links between IGSO satellites and MEO satellites.

At present, academia has carried out some related researches on the link-building of multilayer satellite network. Harathi et al. proposed a link allocation algorithm for the dual-loop satellite communication network, aiming at maximizing the number of links and minimizing the number of link handoffs [2]. Chen et al. proposed a link-building strategy based on maximum link duration, which chooses the object satellite to build links by comparing the duration of coverage of low earth orbit (LEO) satellites by geosynchronous earth orbit (GEO) satellites [3]. Based on the characteristics of space network, to reduce the time for satellite data to be sent to the ground station and the number of ground stations, Zhang designed a link-building allocation scheme [4]. Shi and others considered the queue size, the number of connection nodes, link duration, and also considered the network traffic and other factors when choosing the inter-layer links [5]. The author of this paper has studied the inter-layer topology design for LEO/MEO double-layer constellation [7], and also studied the inter-layer topology design considering beam coverage for IGSO/MEO constellation [8].

At present, in the few studies of multilayer satellite inter-satellite link-building strategies, most of them start with the maximum link duration strategy to construct spatial network topology. The maximum link duration algorithm has the advantage of less link switches [6]. However, for incomplete IGSO/MEO constellation, there is less satellites and less available links, and the constellation configuration is destroyed. So the maximum link duration strategy often causes a few links to occupy the link-building resources for a long time, which results in a network with less node coverage and less links. Therefore, for incomplete IGSO/MEO constellation, the multilayer satellite network constructed with the maximum link duration strategy has poor topological performance.

In this paper, we study network topology optimization design technology for incomplete IGSO/MEO constellation. Firstly, based on link accessibility constraints, the inter-satellite link accessibility is analyzed, including intra-layer links among IGSO satellites, inter-layer links between IGSO and MEO, and intra-layer links among MEO satellites. Then, an inter-satellite link-building method based on dynamic programming is proposed for multilayer satellite network. Finally, this paper simulates the inter-satellite link accessibility and the inter-satellite link-building strategy, and analyses the performance, such as the number of links, the frequency of link switching, and compares the performance with the results of the commonly used maximum link duration strategy.

2 Link Accessibility Analysis

Because incomplete IGSO/MEO constellation network is a multilayer multi-orbit network, its nodes cannot be kept visible to each other at any time, so inter-satellite links need to be switched dynamically, and the topology also changes dynamically, which is specifically reflected in inter-layer links between IGSO satellites and MEO satellites [7], and MEO inter-orbit links, and so on. Therefore, before designing links, the link accessibility among satellites needs to be constrained and analyzed.

The accessibility of inter-satellite links is related to the position of two satellites at both ends of the link, the number of communication terminals carried by satellites and the pitch-azimuth constraints of orbital dynamics. When the orbital elements of a satellite at a given time is known, the right-angular inertial coordinates of a satellite at any time can be calculated. And the number of communication terminals per satellite in IGSO/MEO constellation is limited. So if all terminals of a satellite are occupied, the satellite does not have link accessibility with other satellites. According to the overall configuration and layout of IGSO and MEO satellites, when the communication terminal is in working state, its antenna +Z axis coincides with the satellite body +Z axis, that is, the communication terminal points to the earth center, and the half-angle of the antenna is limited. Figure 1 is a schematic diagram of the relative position of the satellites. Among them, point O denotes the earth's core, line OP is perpendicular to line IM, line OQ is perpendicular to line MM', angle α is half angle of the communication terminal on satellite I, angle β is half angle of the communication terminal on satellite M or satellite M'.

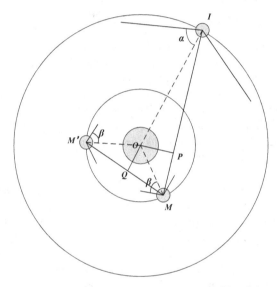

Fig. 1. IGSO/MEO constellation link accessibility diagram

According to the author's previous research [7], the accessibility conditions of the different-layer satellites I and M are as follows:

- Satellite I and M have free communication terminals
- $\angle OMP < \beta$
- $\angle OIP < \alpha$
- $|OP| < R_e + h_{atm}$

R_e is the radius of the earth and h_{atm} is the atmospheric thickness.

Similarly, the accessibility conditions of the same-layer satellites M and M' are as follows:

- Both satellite M and satellite M′ have free communication terminals.
- $\angle OMQ < \beta$
- $\angle OM'Q < \beta$
- $|OQ| < R_e + h_{atm}$

3   Link-Building Strategy of Multilayer Satellite Network Topology Based on Dynamic Programming

Figure 2 shows the composition of multilayer satellite network based on incomplete IGSO/MEO constellation. Its network nodes include IGSO satellites and MEO satellites, etc. This section studies the strategy of inter-satellite network topology design based on incomplete IGSO/MEO constellation. Based on link accessibility and according to different optimization objectives, the appropriate link-building strategies of IGSO intra-layer links, MEO intra-layer intra-orbit links, IGSO and MEO inter-layer links and MEO intra-layer inter-orbit links are studied. A network topology design scheme with fewer link switches, higher transmission bandwidth and wider access user coverage is designed.

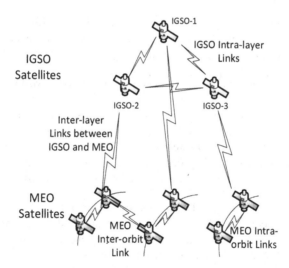

Fig. 2. The node and link composition of incomplete IGSO/MEO constellation

Firstly, we study the strategy for intra-layer links among IGSO satellites and intra-orbit links among MEO satellites. According to the link accessibility analysis in Sect. 2, the two types of links are always accessible under the given accessibility constraints. Therefore, during the whole life cycle, all IGSO satellites are able to build links in pairs, and the MEO satellites in the same orbit are also able to build links with each other. These two types of links are given priority to build links.

Then, we study the strategy for inter-layer links between IGSO and MEO and inter-orbit links among MEO satellites. Because the number of IGSO satellites is small and the number of links between IGSO and MEO is very important to the performance of space information network, this paper takes IGSO satellites as the core to design the links between IGSO and MEO. Based on the designed inter-layer links between IGSO and MEO, we also need to find a link-building strategy for the inter-orbit links in MEO layer when the communication terminals are surplus. Because the number of links and network coverage are very important to the data transmission capability of network topology, the link-building strategy must consider building as many links as possible. In order to reduce the increase of network delay in the process of network topology change, the frequency of inter-satellite link switching should also be reduced. In order to achieve this series of optimization objectives, this paper proposes a dynamic programming based inter-satellite link-building strategy. Dynamic programming is a classical method to solve the optimization problem of multi-stage decision-making process. It can transform the multi-stage process into a series of single-stage problems and solve them one by one by using the relationship between each stage. Inter-satellite link-building design can be regarded as a multi-stage decision-making problem with a chain structure, which can be solved by dynamic programming because it has the characteristics of satisfying the optimization principle, having no aftereffect and including overlapping sub-problems.

Based on dynamic programming, the inter-satellite link-building strategy is computed in the following steps:

Stage Division. The process of inter-satellite link-building is divided into decision-making stages based on time. Each second corresponds to a decision-making stage. $k = 1, 2, \cdots, N$, N is the total number of stages.

State Selection. The link state of each satellite in the IGSO/MEO constellation at this stage is expressed by the link state of each satellite with other satellites. S_k is used to denote the link-building state variable in stage k. For the link-building state $S_k = \{s_{i1}, s_{i2}, \cdots, s_{ix}, \cdots, s_{iu}, s_{m1}, \cdots, s_{my}, \cdots, s_{mv}\}$, s_{ix} is used to denote the set of satellites linked with the IGSO satellite x in stage k, u is used to denote the total number of IGSO satellites, s_{my} is used to denote the set of satellites linked with the MEO satellite y in stage k, and v is the total number of MEO satellites.

Decision-Making. Link-building decision represents the link-building choice between IGSO and MEO layers at stage k, and $u_k(S_k)$ represents the decision variables at stage k when the state is S_k. $D_k(S_k)$ denotes the set of admissible decision-making under the state S_k in stage k. Obviously, there are $u_k(S_k) \in D_k(S_k)$.

Determination of State Transfer Equation. Given the stage k, the link-building state S_k and the decision variable u_k, the stage $k + 1$ will produce the link-building state S_{k+1}, that is, S_{k+1} is determined by S_k and u_k, and the state transition equation $S_{k+1} = T_k(S_k, u_k)$ is obtained.

Determination of Index Function and Optimum Value Function. For the inter-satellite link-building strategy, $V_{k,N}$ is chosen as the index function to represent the link-building loss from stage k to stage N, that is, the weighted sum of the number of unbuilt links and the number of link switches. Link-building loss is chosen as the optimization objective because it is separable and has indicator additivity, i.e.

$V_{k,N} = \sum_{j=k}^{n} v_j(S_j, u_j)$. $v_j(S_j, u_j)$ represents the link-building loss in stage j. Define $v_j(S_j, u_j) = \lambda \cdot p + \eta \cdot q$, where p is the number of links who are accessible in stage k but do not be built, λ is the penalty weight factor for failing to build a link, q is the number of link switches occurring in stage k, and η is the penalty weight factor for link switching. Further, the optimization objective function is defined as $f_k(S_k) = \min_{u_{k,N}} V_{k,N}$. That is, from stage k to stage N, the value of the obtained index function by adopting the optimal strategy, $u_{k,N}$ indicates one strategy from stage k to stage N.

Solution of Inter-satellite Link-Building Problem. Inverse recurrence is the concrete way to realize the optimal strategy of dynamic programming, that is, to deduce from the final state forward, recording the feasible decision-making and the corresponding cost in the previous stage, which to evaluate the overall decision-making cost in the next decision-making. In this way, the final feasible decision-making set can be obtained by inverse recurrence to the initial state. And also choosing the best strategy based on the records, and then tracing back to get the whole decision-making sequence.

Based on the above analysis of dynamic programming process, the recursive equation is as follows:
The stage N:

$$f_N(S_N) = 0$$

The stage k:

$$f_k(S_k) = \min_{u_k}\left[v_k(S_k, u_k) + f_{k+1}(S_{k+1})\right]$$

4   Simulation Experiment and Verification Analysis

In this section, we will simulate and analyze the multilayer satellite network topology link-building strategy based on incomplete IGSO/MEO constellation.

In this paper, the IGSO/MEO constellation is simulated. The constellation consists of three IGSO satellites and 24 MEO satellites. Among them, the IGSO satellite orbital altitude is 35786 km and the orbital inclination is 55°, which are distributed in three orbital planes. RAAN of the three satellites are 120° apart from each other. The ground track of the three satellites coincide. The longitude of the intersection point is 118°E. MEO constellation is Walker 24/3/1 configuration. Orbital altitude is 21528 km. Orbital inclination is 55°. In order to meet the needs of step-by-step construction of IGSO/MEO constellation in the future, this paper pick some satellites from the constellation to form incomplete IGSO/MEO constellation. These satellites include three IGSO satellites, as I1, I2, I3. And they also include six MEO satellites, as M1 to M6, which are selected from three orbits and two satellites per orbit. The simulation time is one week because the constellation orbital period is one week. The simulation step is one second.

In this paper, STK (Systems Tool Kit) is used to simulate the inter-satellite link accessibility in IGSO layer. According to the STK simulation results, the three satellites in the IGSO layer can build links with each other during the simulation time, which ensures the integrity and continuity of the links in the IGSO layer.

STK is also used to simulate the intra-orbit link accessibility of MEO layer. According to STK simulation results, the inter-satellite links of M1 and M2, M3 and M4, M5 and M6 in MEO layer can be built continuously during the simulation time.

In this paper, the inter-layer links between IGSO and MEO are simulated, too.

Figure 3 is a diagram of the link accessibility interval between I1 and all six MEO satellites. The link accessibility of I2 and I3 with six MEO satellites is similar to that of Fig. 3. It can be seen from the Fig. 3 that although the link between MEO satellites and the I1 satellite is not sustainable, if all MEO satellites are considered, the I1 satellite can continuously establish the inter-layer link with the MEO layer at any time, which also lays the foundation for the effectiveness of the inter-layer link design strategy.

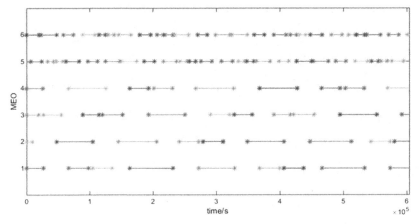

Fig. 3. Link accessibility interval between IGSO1 and all MEO satellites

The inter-orbit link accessibility of MEO layer is also simulated and analyzed. Due to the simulation results, the inter-orbit links in MEO layer can be accessed continuously during the simulation time, but because of the limited satellite communication terminal resources, the actual inter-orbit link-building in MEO layer is subject to the inter-layer link-building between IGSO and MEO.

According to the simulation of inter-satellite link accessibility, the authors use the dynamic programming based inter-satellite link-building strategy to obtain the high-speed network topology of incomplete IGSO/MEO constellation. The weight factor λ in the loss formula $v_j \left(S_j, u_j \right) = \lambda \cdot p + \eta \cdot q$ of phase j is defined as 100 and η is defined as 10, which emphasizes that more inter-satellite links are the primary optimization objective in the process of network topology design, and less link handover is the second optimization objective.

Figure 4 shows the results of inter-layer link-building with MEO from the perspective of IGSO, (a) shows the results of link-building using the maximum link duration strategy, (b) is the results of link-building using the thinking of dynamic programming. The numbers in the figure indicate the serial number of MEO satellites, which are from 1 to 6. As we can see from Fig. 4, using the maximum link during strategy cannot guarantee that each IGSO satellite can establish an inter-layer link with MEO layer satellite at any time, while using dynamic programming to build a link can ensure that each IGSO satellite can build an inter-layer link with one MEO satellite every moment. During the simulation time, using the maximum link duration strategy, it can be built a link between IGSO and MEO by only 94.3% of the time, while using dynamic programming can ensure that three IGSO satellites can establish an inter-layer link in 100% of the time, which is obviously a better multi-layer topology. At the same time, the total number of inter-layer link handoffs during that time is only 96 times, and the less number of link handoffs can make the routing algorithm have enough time to converge, which can greatly improve the data transmission performance of multi-layer spatial networks.

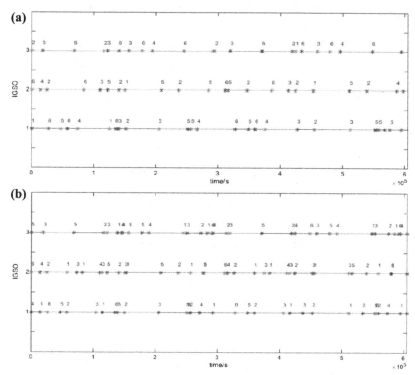

Fig. 4. Inter-layer link handover with MEO from IGSO perspective. (a) Link-building using maximum link time strategy; (b) Link-building using dynamic programming

Figure 5 shows the results of inter-layer link-building with IGSO from the MEO perspective. The numbers in the figure indicate the serial number of IGSO satellites, which are from 1 to 3. Inter-orbit link-building in MEO layer is also simulated. Because the number of IGSO satellites is smaller than that of MEO satellites, a MEO satellite can't link with IGSO satellites in all time. However, because of the intra-orbit and inter-orbit links among MEO satellites, six MEO satellites can be covered by high-speed space networks in most of the time, which is of great significance to improve the coverage of the network and the performance of the network topology.

Figures 6, 7 and Table 1 show the duration attributes of the network topology snapshots built by the strategy introduced in this paper. As can be seen from the chart, the average snapshot duration is about two hours, and most of the snapshots last more than one hour. Topology switching does not occur frequently, which ensures the stability of the network.

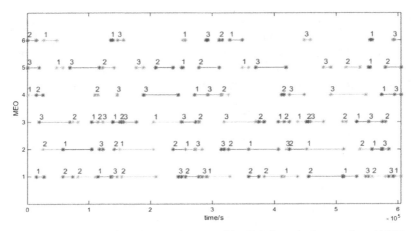

Fig. 5. Inter-layer Link Handover Diagram with MEO from the Perspective of MEO

Table 1. Snapshot duration statistics

Snapshot duration statistics	Time (s)
Minimum duration	13
Maximum duration	28268
Average duration	7286

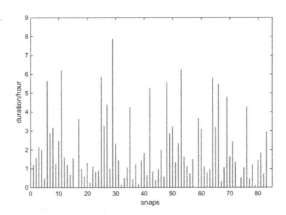

Fig. 6. Inter-satellite link duration distribution

5 Discussion and Future Research

In this paper, the multilayer satellite network topology design strategy is applied to incomplete IGSO/MEO constellation. Future research will focus on the application of the design strategy to the Space-earth integrated network. IGSO/MEO constellation,

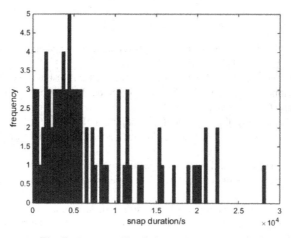

Fig. 7. Inter-satellite link duration histogram

ground stations and accessed users constitute a Space-earth integrated information network. It is of great significance to design the optimal link topology scheme with fewer link switches, higher network transmission bandwidth and wider coverage of accessed users for realizing the networking development of space systems in China. In the future, the network design strategy can be applied to civil projects such as aircraft air networking, real-time ship communications, accessing to LEO Internet satellites, which can implement the strategy of civil-military integration, develop the commercial aerospace industry in depth, and occupy the strategic commanding heights in the emerging field of space-based information services.

6 Conclusion

This paper studies the inter-satellite network topology link-building strategy based on incomplete IGSO/MEO constellation. Firstly, based on the relative position and link constraints of incomplete IGSO/MEO constellation satellites, the accessibility of inter-satellite links is analyzed and the model is given. Then, based on the link accessibility of dynamic network topology and based on different optimization objectives, the appropriate link-building strategies of IGSO intra-layer links, MEO inter-orbit links, IGSO and MEO inter-layer links and MEO intra-orbit links are studied. A dynamic programming based inter-satellite link-building strategy is proposed, with fewer link switches, higher transmission bandwidth and wider accessed user coverage. Finally, this paper simulates the inter-satellite link accessibility and the inter-satellite link-building strategy, and analyses the strategy performance, such as the link numbers, link switching frequency and node coverage. The simulation results show that the proposed dynamic programming based inter-satellite network topology link-building strategy can construct high-speed spatial network topology with superior performance, and is superior to the commonly used maximum link duration strategy in terms of the number of inter-layer links and network coverage.

References

1. Shi, W., et al.: Distributed contact plan design for multi-layer satellite-terrestrial network. China Commun. **15**(01), 23–34 (2018)
2. Harathi, K., Krishna, P., Newman-Wolfe, R.E., Chow, R.Y.C.: A fast link assignment algorithm for satellite communication networks. In: Twelfth Annual International Phoenix Conference on Computers and Communications (1993)
3. Chen, C., Ekici, E., Akyildiz, I.F.: Satellite grouping and routing protocol for LEO/MEO satellite IP networks. In: ACM International Workshop on Wireless Mobile Multimedia. ACM (2002)
4. Zhang, T., Ke, L., Li, J., et al.: Fireworks algorithm for the satellite link scheduling problem in the navigation constellation. In: 2016 IEEE Congress on Evolutionary Computation (CEC) (2016)
5. Shi, W., Gao, D., Zhou, H., et al.: Traffic aware inter-layer contact selection for multi-layer satellite terrestrial network. In: GLOBECOM 2017 – 2017 IEEE Global Communications Conference. IEEE (2017)
6. Wu, T., Wu, Q.: Performance Analysis of the inter-layer inter-satellite link establishment strategies in two-tier LEO/MEO satellite networks. J. Electron. Inf. Technol. **30**(01), 67–71 (2008)
7. Yan, H., Guo, J., Wang, X., Zhang, Y., Sun, Y.: Topology analysis of inter-layer links for LEO/MEO double-layered satellite networks. In: Yu, Q. (ed.) SINC 2017. CCIS, vol. 803, pp. 145–158. Springer, Singapore (2018). https://doi.org/10.1007/978-981-10-7877-4_13
8. Yan, H., Zhang, Y., Zhang, R., Zeng, L., Jia, W.: Inter-layer topology design for IGSO/MEO double-layered satellite network with the consideration of beam coverage. In: 2018 IEEE 18th International Conference on Communication Technology (ICCT), Chongqing, pp. 750–754 (2018)

Capability Assessment of Networking Information-Centric System of Systems: Review and Prospect

Yang Guo, Jiang Cao, Yuan Gao, Yanchang Du[✉], Shaochi Cheng, and Shuang Song

PLA Academy of Military Science, Beijing 10091, China
guoyangnudt@gmail.com, duyanchang198@163.com

Abstract. This paper reviews the research progress of capability assessment of the networking information-centric system of systems. The concept, characteristics and capability structure of the networking information-centric system of systems are summarized, and the challenges of capability assessment are analyzed. For the problem that the nonlinear characteristics of networking information-centric system of systems capabilities are difficult to evaluate, the idea of using machine learning methods to solve this problem is proposed, and the characteristics and applications of these methods are analyzed.

Keywords: Networking information-centric system of systems · Capability assessment · Nonlinear · Artificial intelligence

1 Introduction

The system of systems consists of a series of independent systems and can further generate new capabilities through cooperation between these systems [1]. At present, research on system of systems mainly focus on its concept, design and optimization methods, evolution modeling and simulation, and capability assessment [2]. Through the capability evaluation of the networking information-centric system of systems (NIC SoS), we can find its shortcomings, which is of great significance for its future development and construction. This paper mainly discusses the concept, characteristics, capability structure and capability assessment methods of the SoS.

2 Networking Information-Centric System of Systems and Its Capabilities

2.1 The Concept of Networking Information-Centric System of Systems

The authoritative definition of the system of systems is mostly based on the definition given in the system engineering guide issued by the US Department of Defense in 2008. An SoS is defined as a set or arrangement of systems that results when independent and

© Springer Nature Singapore Pte Ltd. 2020
Q. Yu (Ed.): SINC 2019, CCIS 1169, pp. 39–46, 2020.
https://doi.org/10.1007/978-981-15-3442-3_4

useful systems are integrated into a larger system that delivers unique capabilities [3]. The NIC SoS capability is the ability to perform a series of activities under given conditions to achieve the desired results. Most of the studies limit the capabilities of NIC SoS to the scope of hardware devices and the corresponding functional software. However, we believe that the capabilities of NIC SoS not only include the application systems (hardware and software), but also include advanced concepts, organization structure, operation mechanisms, personnel skills, training level and so on.

2.2 The Characteristics of Networking-Centric System of Systems

The networking information-centric system of systems has the following characteristics:

(1) **Nonlinearity**

Nonlinearity can be either a sudden increase in ability or a collapse in ability. For example, the space information networks use satellite, stratospheric airships, unmanned aerial vehicles and other platforms to construct a SoS that can acquire, transmit and process information in real time. It extends human activities to all the domains and even deep space so that human cognition ability has reached to a new level. This is not a linear change, but like a quantum transition from one energy level to another. Another example, the emergence of UAV swarming technology may lead to the collapse of traditional air defense systems. In the past, the air defense system was effective for conventional aircraft. However, after the emergence of UAV swarming technology, the interception capability of the target would be greatly reduced or even completely collapsed.

(2) **Collaborative**

The capabilities of NIC SoS may be the ability of a single system within the NIC SoS, or the new capabilities provided by the systems working together. And these new capabilities are not available in a single system. The unique capability is not just a simple combination of the functions of the various systems. It is achieved through interaction and collaboration between these systems.

(3) **Autonomy**

The NIC SoS does not necessarily have a unified control mode, and the its components can operate, manage and control independently according to their own characteristics. Each subsystem is an independent system with independent capabilities and corresponding functions. These independent systems are the basis of the NIC SoS capabilities.

(4) **Evolutionary**

The NIC SoS will evolve and grow gradually. For the space information networks, with the introduction of advanced concepts, the development of technology, the adjustment of operation mechanisms and the cultivation of talents, the SoS capabilities will change accordingly. This is a process of continuous evolution.

2.3 The Capability Framework of Networking Information-Centric System of Systems

The construction of the NIC SoS itself is hierarchical, and its capability is also aggregated according to the hierarchical relationship of the NIC SoS. According to the NIC SoS capability aggregation process, its capability can be divided into three levels, namely the fundamental layer, the domain capability layer and the SoS capability layer. The bottom layer is the fundamental layer, which includes not only the application system, but also the concepts, organization structure, operation mechanism, personnel skills and so on. The middle layer is the domain capability layer, which refers to the division of the NIC SoS into different functional domains. The function integration of different systems is realized according to the logical relationship and characteristics of the domain, and the ability with domain features across systems is formed. The top level is the NIC SoS capability layer, which refers to the ability to integrate various domains according to the needs of the task. The hierarchical structure of the NIC SoS capabilities is shown in Fig. 1.

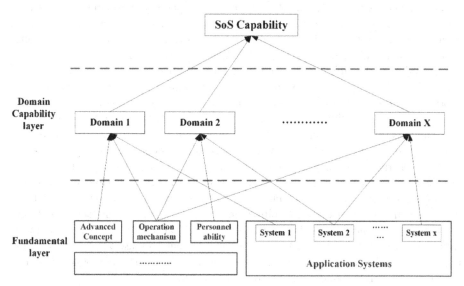

Fig. 1. SoS capability framework

3 Networking Information-Centric System of Systems Capability Assessment

3.1 Classification

(1) **Static assessment**

The NIC SoS capabilities are inherent and static attributes of the SoS. They are related to the function, quantity and structure of the member system, and have nothing to do with the specific activity process [2].

(2) **Dynamic assessment**

The NIC SoS capabilities are relative, especially for confrontation scenarios. Relativity means that the SoS ability is generated in the movement. For different environments and opponents, the SoS capability is different.

The focuses of the above two assessment ideas are different. Static assessment can evaluate the construction and operation of the SoS itself. The goal of this kind of assessment is the absolute ability that the SoS can achieve under ideal condition. For example, the US military has proposed a two-hour global strike capability requirement, which is an assessment of the absolute capabilities of its own SoS. The difficulty of static assessment is relatively small because it is easier to get the information needed for the assessment.

Dynamic assessment is more suitable for confrontation scenario because it reflects the comparison of the capabilities between opponents. The US military has long recognized this problem and pointed out that combat effectiveness is not a numerical value. It can be evaluated, but it cannot be quantified because the combat effectiveness is always relative. Combat effectiveness makes sense only when the environment and the opponent are clear. Dynamic assessment needs to evaluate the capability of SoS of both sides, and take into account the complex confrontation between the two SoS at the same time. Therefore, considering the difficulty of establishing evaluation methods and obtaining useful information, it is much more difficult than static assessment.

According to the difficulty of obtaining useful information, the dynamic assessment can be divided into two situations. One is the ability assessment between opponents in sports competitions. The information on both sides is more transparent and easier to conduct assessment activities. The other is the military confrontation scene. Dynamic assessment is difficult to perform because it is difficult to get details of the opponent. The information needed includes the performance of the opponent's weapons, combat methods, operation mechanisms, personnel skills and so on. Therefore, at this stage, the US military mainly uses static assessment ideas for research. However, with the development of the concept of warfare, the future military confrontation will no longer be a competition for absolute capabilities in certain aspects, such as tactical fighting between fighters and fighters, but more emphasis on confrontation between the SoSs. Both sides will look for the weaknesses of the opponent SoS, use their own strengths to attack enemy's weaknesses, and achieve the goal of defeating the opponent. But for now, there are still few studies on dynamic assessment.

3.2 Assessment Method

There are mainly two kinds of methods for evaluating the SoS capability: (1) Mathematical analytical method, which is a method for calculating the ability value based on the functional relationship between the ability and the given condition, such as analytic hierarchy process (AHP). At present, most mathematical analytical methods use a linear method in the final calculation of SoS capability, which cannot solve the problem of nonlinearity of SoS capability (2) The simulation method refers to the method of using the modeling and simulation technology to conduct a large number of experiments and evaluating the system capability through analyzing of the simulation data. This method can flexibly adjust the influencing factors and can fully evaluate the SoS capabilities in various scenarios. The main problem faced by this method is the "combination explosion" problem caused by SoS uncertainty. However, with the improvement of computer computing power and the progress of artificial intelligence technology in the future, simulation method will play a greater role.

4 Challenges

There are two difficulties in the study of SoS capability assessment.

(1) The role of human beings in the SoS is difficult to assess. Although AI technology is developing vigorously, the existing technology is limited to enable machines to operate under the rules set by human beings to achieve a higher degree of automation. Artificial intelligence technology has not yet achieved human-like creativity. From the design to the operation management of the SoS, human beings need to play a central role, especially in the confrontation scenario. However, it is difficult to quantitatively evaluate the role of people in SoS design and operation management, which also brings challenges to SoS evaluation.

(2) There is a lack of methods for evaluating the nonlinear characteristics of SoS capability. As mentioned above, most of the mathematical analytic methods use linear weighting method in the synthesis of capability indicators, which can't reflect the nonlinear characteristics of SoS capability. Although the simulation evaluation method can flexibly adjust the influencing factors to reflect the change of SoS capability, it still can't explain the internal relationship between the SoS factors leading to the transition of SoS capability. This nonlinearity also makes it difficult to evaluate the SoS capability.

5 Prospect

In view of the nonlinear characteristics of the SoS capability, it is believed that research should be carried out in two steps. The first step is to study whether the SoS capability has jumped to another level when the environments and the SoS have changed. Different levels of SoS capability should be identified from the appearance. On this basis, the second step is to study the root causes and scientific laws that lead to the transition of the SoS capability, helping us to better build and use the SoS.

5.1 Research on the Expression of Nonlinear Characteristics

(1) **Using Logistic Regression Method to Describe the Nonlinear Characteristics of SoS Capabilities**
Regression analysis is an effective method of using data to analyze the relationship between elements and results. Due to the nonlinearity of the SoS capabilities, commonly used linear regression models are not applicable. The function we want should be to accept all the inputs and predict the capability level. The heaviside step

function has this property. However, the problem with the heaviside step function is that the function jumps from 0 to 1 at the jump point, which is mathematically difficult to handle. The logistic regression model (LRM) uses the sigmoid function, which outputs values ranging from 0–1 and is very sensitive at critical values (shown in Fig. 2). Its effect is similar to the heaviside step function, but it is mathematically easier to handle. This method can effectively distinguish the SoS capabilities of different levels, meeting the requirements of exhibiting the nonlinear changes in SoS capabilities [4].

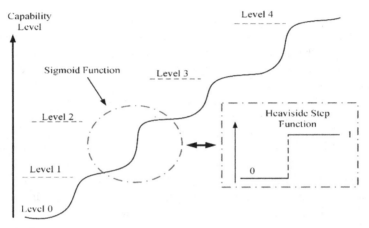

Fig. 2. Capability level transition represented by logical regression method

(2) **Using Support Vector Machine Method to Describe the Nonlinear Characteristics of SoS Capabilities**

There are many dimensions in the evaluation of SoS capability, and the factors of different dimensions can be combined into complex nonlinear problems. For the nonlinear model, we can try to use the nonlinear support vector machine (SVM) to describe the nonlinear characteristics of the SoS capability. When the nonlinear problem is difficult to solve in low dimension, it can be mapped from the low dimension space to another high dimension space. After space transformation, the low-dimensional nonlinear problem can be transformed into the high-dimensional linear problem, so as to solve the problem easily [5]. The separating planes that distinguish the different capability levels of the SoS are called hyperplane. In the two-dimensional case, the separating hyperplane is a straight line. In the three-dimensional case, the separating hyperplane is a plane (shown in Fig. 3). In higher dimensional case, the separating hyperplane is the decision boundary of classification. By using the support vector machines for classification, different levels of SoS capability can be distinguished by different regions in high-dimensional space, and the nonlinearity of SoS capability can be described.

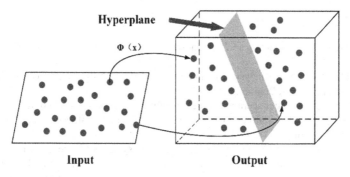

Fig. 3. Capability level transition represented by support vector machine method

5.2 Research on the Main Influencing Factor of SoS Capabilities and the Interaction Relationship Between Them

The above two machine learning methods can display the nonlinear characteristics of SoS capability through graphics or other ways. They only solve the problem of exhibiting nonlinear characteristics, and there are other ways to achieve the same effect. However, these methods do not reveal the fundamental reasons leading to the transition of SoS capability, that is, the main influencing factors of SoS capability and the interaction between these factors. At present, most of the SoS capability indicators are designed according to our own experience. However, due to the limitations of human cognition, on the one hand, there may be hidden indicators that have a greater impact on the SoS capability has not been found; on the other hand, there is insufficient analysis of the implicit relationship between the existing indicators. Big data and artificial intelligence technology provide us with a good opportunity to explore hidden indicators or implicit relationships that we did not think of or beyond the scope of human cognition. This is the core of understanding the nonlinear characteristics of the SoS capabilities.

For example, generative adversarial net (GAN) is a promising technology [6]. Traditional machine learning methods usually define a model for data to learn. Suppose that the original data belongs to a gaussian distribution, but its parameters are unknown. We could define the gaussian distribution and then use the data to learn its parameters to get the final model. Another example, we can define a classifier (such as SVM) and then perform various mappings on the data, turning it into a simple distribution problem. All these methods directly or indirectly tell the data how to map. The only difference is that the different mapping methods have different effects. However, whether the predefined mapping relationship can represent the real mapping relationship is unknown. The GAN is different. The trained generative model can generate a real sample (such as a face image) through noise, indicating that the generative model has mastered the real mapping relationship from random noise to face image. However, the obtained mapping relationship is obviously unknown at first. GAN can learn the true mapping relationship between data and sample set. It is a valuable method for us to get the true mapping relationship between the SoS indicators and its capabilities.

References

1. Si, G.Y., Wang, Y.Z.: Modeling and Simulation of Cyberspace Warfare. Science Press, Beijing (2019)
2. Zhang, T.T.: Evolution Analysis Method of Networking Information-Centric System of Systems. Science Press, Beijing (2018)
3. Office of the deputy under secretary of defense for acquisition and technology, systems and software engineering. Systems engineering guide for system of systems, version 1.0. ODUSD (A&T) SSE, Washington, DC (2008)
4. Zhang, X.C., Ma, Y.H.: Capability model of combat system of systems and measurement method of armament contribution to combat system of systems. Syst. Eng. Electron. **41**(4), 843–849 (2019)
5. Harrington, P.: Machine Learning in Action. Manning Publications, Greenwich (2012)
6. Goodfellow, J.I., Pouget-Abadie, J., Mirza, M., et al.: Generative adversarial nets [EB/OL]. https://arxiv.org/abs/1406.2661

AI Based Supercomputer: Opportunities and Challenges

Jiang Yujuan[1], Li Xiangyang[2(✉)], and An Binlai[3]

[1] AECC Sichuan Gas Turbine Establishment, Chengdu 610000, Sichuan, China
[2] Academy of Military Science of the PLA, Beijing 100142, China
lixyljx@163.com
[3] UNIT 37100, Huludao 610000, Liaoning, China

Abstract. Since 2013, China's supercomputer has been ranked first in the global supercomputer rankings, and now the United States has returned to its peak, engineers at the US Department of Energy's Oak Ridge National Laboratory released "Summit," a supercomputer with powerful performance that surpasses the current record holder: China's Shenwei·Taihu Light. AI has played an important role in recent development of supercomputer systems. In this paper, we discuss the opportunities and challenges of future supercomputer systems based on the thinking of AI.

Keywords: Supercomputer · AI · GPU

1 Introduction

The Oak Ridge National Laboratory in eastern Tennessee announced the development of the new Supercomputer Summit [1], the most powerful computing machine on the planet. It is designed to extend artificial intelligence technology in part.

Supercomputers are slightly eclipsed by the era of cloud computing and big data centers. But many tricky computing problems still require large machines [2].

A US government report in the past year said that the United States should invest more in supercomputing to catch up with China in defense projects such as nuclear weapons and hypersonic aircraft, and commercial innovation in the aerospace industry [3], oil exploration and pharmaceutical industries. The super-calculated Summit built by IBM is the size of two tennis courts, and its circulatory system consumes 4,000 gallons of water per minute to cool 37,000 processors. According to Oak Ridge Labs [4], the results of the standard metrics used to evaluate supercomputers show that the peak performance of the new machine can reach 200 teraflops per second, or 200 petaflops. The speed is about one million times that of a typical laptop, and the peak performance is almost twice that of the previous super-powered Shenwei·Taihu Lake [3].

In an early test, researchers at Oak Ridge National Laboratory used Summit to perform more than one million mega-calculations per second in a project that analyzed differences in human genetic sequences [5]. They called this the first scientific calculation to meet the computational scale requirements [6]. The best supercomputer in the United

© Springer Nature Singapore Pte Ltd. 2020
Q. Yu (Ed.): SINC 2019, CCIS 1169, pp. 47–55, 2020.
https://doi.org/10.1007/978-981-15-3442-3_5

States not only has a great influence on the geopolitics of computing power, but its design is also more suitable for running the popular machine learning technology in technology companies.

In recent years, one of the reasons for computer science to achieve breakthroughs in speech recognition and games is that researchers have found that graphics chips can provide greater power to deep neural networks [7], a machine learning technology.

According to reports, the new Summit uses 27,648 new NVIDIA Volta GPUs with Tensor Core and more than 9000 traditional processors from IBM, including Power 9 [8]. With the help of high-speed interconnect technology such as NVLink, Summit deployed six GPUs per node, which enabled its analog performance to be ten times that of the previous Titan, with 95% of its computing power coming from the GPU.

In this paper, we discuss the AI based HPC systems, the rest of the paper is organized as follows: in Sect. 2, we discuss the combination of AI and HPC systems, in Sect. 3, we discuss the opportunities and challenges of HPC, in Sect. 4, conclusion is given as summary.

2 AI Based Supercomputer: The Future

In September 2019, the National Science and Technology Council (NSTC) released the "High Performance Computing, Big Data and Machine Learning Integration" report for the US Network and Information Technology Research and Development Program (NITRD) big data and high-end computing research and development agencies The group summarized the meeting of the same name held in October last year.

(1) With the rapid increase in the amount of data, high-performance computing (HPC), big data (BD) and machine learning (ML) are merging under the impetus of scientific demand. The generation of data is no longer a bottleneck, but instead it is the management, analysis and reasoning of data.

(2) As the contribution of semiconductor scaling to performance improvement is gradually reduced, the heterogeneity of future systems will continue to increase. The system will need to increase overall flexibility and low latency to support new applications more effectively. In addition, because of the current scarcity of data, new tools and benchmarks are needed to address common issues encountered in HPC simulation, big data, and machine learning applications.

(3) The future computing ecosystem will be different from the current computing ecosystem, and more likely to combine edge computing, cloud computing and high performance computing. To achieve this seamless ecosystem, new programming algorithms, language compilers, operating systems, and runtime systems will be needed to provide new abstractions and services. The importance of "edge intelligence computing" is expected to increase, involving intelligent data collection or data classification at the edge of the network (near data sources).

(4) More cooperation between HPC, BD and ML communities is needed to achieve higher school rapid ecosystem development and serve these three types of communities more effectively. The convergence of data analysis and HPC simulation has

made some progress, and due to technical and organizational differences, the software ecosystem supporting the HPC and BD communities presents a completely different situation.

On April 1, 2019, the US Department of Energy (DOE) and Lawrence Livermore National Laboratory (LLNL) jointly announced that they will begin to re-adjust the "Using High Performance Computing for Energy Innovation" (HPC4EI) program in the spring of 2019. The program is led by the US Department of Energy's Lawrence Livermore Laboratory in collaboration with other national laboratories to provide industry with high-performance computing (HPC) expertise from the US Department of Energy's national laboratories. Technology and resources reduce the risk of industrial use of HPC resources and expand the application of HPC in the technology development process. The program will focus on industrial projects that can further save energy and reduce costs. Through the smooth implementation of the plan, the United States hopes to comprehensively improve the research and development potential of the national industrial system in materials, manufacturing, transportation and mobile systems, and further realize the long-term goal of saving energy and reducing costs, and the ability of the United States in the fields of energy, computing and industry. Leading the global boost. Conceptually, HPC is a term in the computer field that refers to computing systems and environments that typically use many processors (as part of a single machine) or several computers organized in a cluster (operating as a single computing resource). Applications running on high-performance clusters generally use parallel algorithms. The simple understanding is to break down a "big problem" into a number of "small problems" according to certain rules, and perform calculations on different nodes in the cluster. The result of the "problem" can be combined into the final result of the "big problem." Since the calculation process when dealing with "small problems" can be done in parallel, the processing time of "big problems" can be greatly shortened. According to a study by the American Air Force Association's Michel Aerospace Strength Institute, in the past few decades, due to the high risk, large investment, and long cycle of the aviation industry, countries and regions represented by the United States and Europe have adopted public The financial investment in this field has been gradually reduced, and the industry has become more conservative. Technologies such as CFD, new materials, and new processes have slowed down in engineering applications, resulting in the number of new aircraft designs and the number of first flights actually decreasing year by year. In addition, the regulatory authorities have continuous high standards for energy consumption, emissions, noise, etc. in the future development of the industry. The traditional design methods and technical potential have been tapped to the limit, and the "experience" approach is increasingly It's hard to continue, and it's getting harder and harder to get started quickly. The emergence of HPC can use supercomputers to push people who are too big (such as stars, galaxies), too small (such as atomic, nanoscale), too fast (such as nuclear fusion), too slow (such as cosmology), Research on issues such as too dangerous/expensive (such as destructive tests). HPC's ability to dismantle complex problems will significantly shorten the time to break through the bottleneck of innovation and solve specific problems, create opportunities for faster realization of technological innovation, and lay the foundation for leap-forward development in the industrial sector.

Therefore, on July 29, 2015, the then US President Barack Obama issued an administrative order to officially launch the National Strategic Computing Program (NSCI), which aims to maximize the development and deployment of high-performance computing (HPC) for economic competition. Scientific discovery. The executive order pointed out that the continuous development and deployment of new computing systems over the past 60 years has enabled the United States to lead the world in computing. In order to maintain and expand this advantage in the coming decades, maximizing the benefits generated by HPC, meeting the growing demand for computing power, and better responding to emerging computing challenges and opportunities, it is necessary to develop and deploy HPC at the national level. Establish a coordinated joint strategic plan. The launch of NSCI will establish a unified, multi-sectoral strategic vision and federal investment plan for the United States, and plan to maximize the benefits of HPC through the cooperation of production, learning and research.

The primary guiding principle emphasized in the NSCI program is the need to widely deploy and apply emerging HPC technologies to maintain the US's leadership in economic competition and scientific discovery. Since the NSCI program is positioned at the national level, this strategy also marks the United States' view of HPC capabilities as the basis for future technological developments in science, technology, military, and industry. As one of the leaders of this program, the US Department of Energy has the HPC capabilities of most of the state-level supercomputers in the United States, and the HPC4EI program came into being in this context.

Intel announced that it will build the first super-computer with billions of floating-point operations per second with high-performance manufacturer Cray at the Argonne National Laboratory under the US Department of Energy, dedicated to traditional high-performance computing and artificial intelligence (AI) design [9].

Just a week before March 12, Nvidia announced a $6.9 billion acquisition of Israeli company Mellanox, a chip maker known for its high-performance computing and networking technologies. Nvidia is aiming to secure the data center through this acquisition.

Not only the technology giants such as Intel and NVIDIA are super-calculated, but artificial intelligence-creating enterprises have also created super-calculations. The computational cluster of the Shangtang Supercomputing Platform [10] has been equipped with more than 14,000 GPUs, with a peak calculation of 1.6 billion times per second, while the national "Taihu Light" peak calculation is only 1.25 billion times per second.

When despising the announcement of the completion of the C round of financing last year, despise has built a very large super-calculation platform in several places in China [11], and the future computing power needs more. The computing power is just like the storage of the year. No matter how fast the expansion is, it will be consumed and it needs to be continuously invested.

In fact, these artificial intelligence companies can fully adopt the cloud service model, such as high-performance computing services such as Leasing Alibaba Cloud, Tencent Cloud, AWS, and Zhongke Shuguang.

2.1 Fast Iteration

Computational forces and algorithms are a set of optimal CPs. If the computational forces and algorithms are developed by themselves, a "chemical reaction" of 1 + 1 greater than 2 is produced. Because both the original algorithm and the computational power use a uniform interface, it is easier to match and coordinate with each other. And data collection, labeling, model building, model training to output SDK every step, can be standardized and automated, the entire chain will run faster, algorithm iteration faster [12]. The super-calculation of public clouds is difficult to match the matching of each enterprise algorithm.

Especially in the face of new demands, such as the need for 1000 GPU card joint training, Alibaba Cloud, Tencent Cloud and other cloud platforms do not have such services, then the new demand cannot continue. In the long run, self-built super-calculation is more conducive to exploring new business. Not long ago, the government of China broke out that AI customer service hit more than 4 billion harassing calls a year [13], and the phenomenon of criminals stealing user consumption information through free public WIFI caused a hot discussion. Behind it is the weak mapping of China's data security protection. If the model training is done through the public cloud platform, the user data can be seen in the cloud platform in theory. Once the data is leaked, it will be an irreversible blow to the user company. At present, 5G commercialization is approaching, and the production mode will be revolutionized in the 5G era [14]. Many terminal data processing can be run in the cloud. This is one of the reasons why Intel, NVIDIA and other giant companies have recently tried to build a super-computing platform, because super-computing is a non-negligible aspect of the 5G era.

Frozen is not a cold day, building a supercomputer is not just a stack of thousands or tens of thousands of GPUs, but also a powerful "management system" - just like the Microsoft Windows operating system. For example, Ali spent many years to create a "pangu distributed system", which became Alibaba Cloud's Windows [15]. Therefore, for artificial intelligence companies, it is necessary to accumulate over-experience in advance.

From the perspective of capital, in the industry environment of the 5G outbreak, the value of super-calculation has become more prominent, and self-built super-calculation has more imagination. In the case of meeting their own computing needs, they can also lease out services to sell to SMEs, and perhaps profitable. It is also a business model.

2.2 Cost Efficiency

Comparison of the cost of self-built computers and leases from AWS. One GPU version is 4–10 times cheaper, and the four GPU versions are 9–21 times cheaper, depending on utilization. AWS pricing includes a discount for three-year and three-year leases (35%, 60%). Assuming a power consumption of $0.20/kWh, one GPU machine consumes 1 kW/hour, and four GPU machines consume 2 kW/hour. Depreciation is conservatively estimated to be linear loss over 3 years. $700 per GPU.

If a company want to use the 2080TI for deep learning computer, you will get an extra $500, and for a 1 GPU machine, it's still 4–9 times cheaper. The reason for this huge cost difference is that Amazon Web Services EC2 (or Google Cloud or Microsoft

Azure) has a GPU price of $3/hour or about $2100/month. Even when you shut down your machine, you still need to pay for the machine at a price of $0.10 per GB per month.

For a $3,000 GPU machine learning computer (1 kW/hour), if you use it often, it will break even within 2 months. Not to mention that your computer is still owned by you, and it has not depreciated much in two months. Again, the 4 GPU version (2 kW/hr) is more advantageous because you will break even in less than a month. (assuming the cost of electricity is $0.20/kWh)

And GPU performance is comparable to AWS. Compared to the Nvidia v100 GPU with next-generation Volta technology, your $700 Nvidia 1080-TI runs 90% faster. This is because there is IO, so even if V100 is theoretically 1.5–2 times faster, IO will slow down in practice. Since you are using M.2 SSD, IO runs very fast on our own computer.

3 Opportunities and Challenges

The advancement of computing power, algorithms and big data is the three foundations of AI development. Alpha Go has defeated the top human Go players in succession. On the one hand, it has benefited from the breakthrough of Monte Carlo algorithm, but the performance improvement of AI server and tens of thousands of players playing big data are also indispensable elements. Facts have proved that the three major elements of computing power, algorithms and big data, artificial intelligence are interdependent and mutually restrictive. If the computing power is not enough, then there is more data that cannot be effectively driven and utilized; if the algorithm is stagnant, then when faced with multiple data levels, the existing computing power will not work; if not large enough and associated The data,

It's like a super sports car with excellent performance and gearbox, but without fuel, the real power of AI can't be played. Whether it is the decryption of human genetic code, or the calculation of space astrophysics, or the use of meteorological cloud maps for accurate weather and disaster prediction… The computational bottleneck encountered in a large number of applications, allowing humans to constantly explore HPC performance limits. Currently in the top TOP 500, the Shenwei·Taihu Light Supercomputer has a peak performance of 125.436 PFlops, but if you want to meet the requirements of the E-level calculation, you need to improve the performance ten times in the light of Shenwei·Taihu Lake. Today, when the law is about to fail, HPC needs to cope with the ever-increasing power consumption challenge while pursuing performance. In terms of power optimization, AI can make a big difference.

AlphaGo's algorithmic results in deep learning were used by Google in a pilot data center for energy optimization, resulting in a 40% reduction in energy consumption in the data center, which greatly enhanced people's confidence in using AI to save energy in large data centers. HPC's powerful computing power can make AI plug in the wings of computing power, and AI will in turn help HPC to achieve more optimized resource allocation and energy management. Therefore, in the intelligent era, HPC and AI become a pair of energy. A good CP that complements and helps each other.

The first big issue is high-bandwidth memory technology. The biggest difference between the supercomputer and the CPU used in our ordinary PC is that the former especially enhances the floating point calculation performance. This year's new generation of Xeon Phi and next year's next-generation Tesla will have 3T Flops double-precision floating-point operation speed, about 12 times that of the mainstream i7 4770 CPU. The Shenwei multicore computing chip developed in China can achieve the performance of 1T Flops, but it will face the bottleneck of memory bandwidth when it continues to improve the index.

Most of the programs that run overtime are more dependent on memory bandwidth. If the bandwidth is insufficient, the floating-point performance indicator is no longer valuable. The new generation of Xeon Phi and Tesla have a memory bandwidth of more than 600 G/s, which is more than 20 times that of mainstream CPUs. In order to achieve such high bandwidth, Intel and Nvidia use memory 3D packaging technology to closely connect memory chips and computing chips. This technology requires in-depth cooperation between processor R&D companies and memory companies, making it far more difficult than traditional memory solutions.

The domestic Shenwei processor and Feiteng processor began research on 3D package memory technology a few years ago. However, due to lack of experience and cooperation with memory companies, the gap between Intel and Intel is still huge. In fact, domestic processors have used memory controller modules sold by third parties in the past, even if there is no experience in the development of traditional memory systems, let alone a new generation of 3D package memory. Intel, Nvidia, AMD and other companies that master 3D memory technology are not likely to sell the corresponding technology licenses to domestic companies. It is difficult for Samsung and other memory manufacturers to provide strong support to domestic enterprises. The memory problem is not solved. Shenwei, Feiteng and other super-calculation chips are difficult to achieve the high computing index of Xeon Phi and Tesla in the same period, so that the domestic super-computing will not be able to face the American opponents equipped with Intel and Nvidia chips.

Another key technology that constrains the development of supercomputers is the high-bandwidth interconnect bus. The PCIe bus that is common in our PCs is too slow for the interconnection of super floating processors with high floating point performance, and the bandwidth of 32 G/s is like a narrow two-lane road. The next generation of Xeon Phi and Tesla will be upgraded to a dedicated bus with a bandwidth of more than 100 G/s, greatly reducing the congestion of data exchange between a large number of processors in the system. Intel will even use silicon photonics for the first time, replacing optical circuits with high-speed information exchange between chips, reducing bus complexity, power consumption, and performance.

Domestic processors are still in a significantly backward situation in this respect. The interconnection scheme used by the new generation of Shenwei, Feiteng and other floating-point processors is still the level of PCIe, and it is difficult to catch up with US companies in two or three years. In the advanced technology field such as silicon photon transmission, domestic enterprises are still in the mid-term research stage, and have a long way to go from actual deployment. Insufficient bus bandwidth means that it is difficult for domestic supercomputers to take the initiative by occupying a larger number of chips, and it is impossible to erase the disadvantages caused by insufficient performance of individual chips.

Imported chips can't be bought in advance, and domestic processors are facing two technical problems that can't match the US products. In the next few years, the chances of China's summit will be very slim. However, the leaderboard is just an honor. It doesn't make much sense to simply pursue the ranking. If China increases its investment in independent chips after the US embargo and prepares for a longer-term future, then after a few years of decline, Chinese supercomputers equipped with domestic processors can still compete with the top systems in the United States. Under. Moreover, the opportunity to make domestically produced processors can make great progress in other fields, and even replace imported processors in some key industries. This is a great opportunity. Perhaps China's self-developed processor will rise from the US embargo policy. The big dream of China's technology industry for many years is now hopeful.

4 Conclusion and Future Work

In terms of high-performance computing capabilities, deep learning requires high-capacity, high-bandwidth parallel storage, high-bandwidth, low-latency interconnected networks, requiring larger GPU clusters, and specialized neural network chips.

In terms of offline platforms, there are mainly X86 CPU peer parallel computing and GPU/MIC heterogeneous parallel computing. Because the amount of data involved in offline training is very large, it is often able to reach the PB level, and the calculation and communication are very dense, because algorithms such as deep neural network (DNN), cyclic neural network (RNN), and convolutional neural network (CNN) are often scalable. Not very high, you need to perform efficient calculations within the node. On the online platform, there are X86 CPU isomorphic parallel computing, GPU/MIC heterogeneous parallel computing, and FPGA heterogeneous parallel computing.

The actual combination of supercomputer and artificial intelligence will lead the future. China has a self-developed super-computing and achieved the fastest in the world, but it is not yet full of "full blood" applications and services for the outbreak of various projects, because the entire software ecology and development ecology still needs to be built.

References

1. Sykes, E.R., Mirkovic, A.: A fully parallel and scalable implementation of a Hopfield neural network on the SHARC-net supercomputer. In: 19th International Symposium on High Performance Computing Systems and Applications (HPCS 2005), Guelph, ON, Canada, pp. 103–109 (2005)
2. Miyano, S.: Revolutionizing cancer genomic medicine by AI and supercomputer with big data. In: 2018 IEEE Symposium on VLSI Technology, Honolulu, HI, pp. 7–11 (2018)
3. Meek, T., Barham, H., Beltaif, N., Kaadoor, A., Akhter, T.: Managing the ethical and risk implications of rapid advances in artificial intelligence: a literature review. In: 2016 Portland International Conference on Management of Engineering and Technology (PICMET), Honolulu, HI, pp. 682–693 (2016)
4. Kepner, J., et al.: TabulaROSA: tabular operating system architecture for massively parallel heterogeneous compute engines. In: 2018 IEEE High Performance Extreme Computing Conference (HPEC), Waltham, MA, pp. 1–8 (2018)

5. Keinan, S., Frush, E.H., Shipman, W.J.: Leveraging cloud computing for in-silico drug design using the quantum molecular design (QMD) framework. Comput. Sci. Eng. **20**(4), 66–73 (2018)
6. Huang, Y., Wang, W., Wang, L., Tan, T.: Conditional high-order Boltzmann machines for supervised relation learning. IEEE Trans. Image Process. **26**(9), 4297–4310 (2017)
7. Huang, Y., Wang, W., Wang, L.: Video super-resolution via bidirectional recurrent convolutional networks. IEEE Trans. Pattern Anal. Mach. Intell. **40**(4), 1015–1028 (2018)
8. Barnell, M., Raymond, C., Capraro, C., Isereau, D., Cicotta, C., Stokes, N.: High-performance computing (HPC) and machine learning demonstrated in flight using Agile Condor®. In: 2018 IEEE High Performance extreme Computing Conference (HPEC), Waltham, MA, pp. 1–4 (2018)
9. Kahle, J.A., Moreno, J., Dreps, D.: 2.1 summit and sierra: designing AI/HPC supercomputers. In: 2019 IEEE International Solid- State Circuits Conference - (ISSCC), San Francisco, CA, USA, pp. 42–43 (2019)
10. Hu, J.R., Chen, J., Liew, B., Wang, Y., Shen, L., Cong, L.: Systematic co-optimization from chip design, process technology to systems for GPU AI chip. In: 2018 International Symposium on VLSI Technology, Systems and Application (VLSI-TSA), Hsinchu, pp. 1–2 (2018)
11. Hu, J.R., Chen, J., Liew, B., Wang, Y., Shen, L., Cong, L.: Systematic co-optimization from chip design, process technology to systems for GPU AI chip. In: 2018 International Symposium on VLSI Design, Automation and Test (VLSI-DAT), Hsinchu, pp. 1–2 (2018)
12. Schaefers, L., Platzner, M.: Distributed Monte Carlo tree search: a novel technique and its application to computer go. IEEE Trans. Comput. Intell. AI Games **7**(4), 361–374 (2015)
13. Codreanu, V., Podareanu, D., Saletore, V.: Large minibatch training on supercomputers with improved accuracy and reduced time to train. In: 2018 IEEE/ACM Machine Learning in HPC Environments (MLHPC), Dallas, TX, USA, pp. 67–76 (2018)
14. IEEE 802 Nendica Report: The Lossless Network for Data Centers: IEEE 802 Nendica Report: The Lossless Network for Data Centers, pp. 1–29, 17 August 2018
15. Lee, J., Lee, C.Y., Kim, C., Kalchuri, S.: Micro bump system for 2nd generation silicon interposer with GPU and high bandwidth memory (HBM) concurrent integration. In: 2018 IEEE 68th Electronic Components and Technology Conference (ECTC), San Diego, CA, pp. 607–612 (2018)

A Semi-physical Simulation Platform Using SDN and NFV for LEO-Based IoT Network

Qianyu Ji and Jian Wang[✉]

School of Electronic Science and Engineering, Nanjing University, Nanjing 210023, China
wangjnju@nju.edu.cn

Abstract. Internet of Things (IoT) is rapidly gaining ground in the scenario of modern wireless communications. The satellite system, which plays a significant role in the development of IoT, is expected to cooperate with the terrestrial components to provide a complementary service. In order to better study the combination of satellite and terrestrial IoT networks, building a simulation environment based on software defined network (SDN) and network function virtualization (NFV) technology to create a real and reliable large-scale test environment is of great significance. This paper provides an architecture of IoT network in a combined low earth orbit (LEO) satellite-terrestrial structure embracing SDN technologies. A semi-physical simulation platform utilizing SDN and NFV for satellite-based IoT Network is demonstrated. Last but not least, two use cases of this platform are discussed. Simulations of satellite routing algorithms and combined LEO-terrestrial mobile network prove the validity and authenticity of the platform.

Keywords: Internet of Things · LEO constellation · Satellite · SDN · Network simulation

1 Introduction

In the context of next generation 5G networks, the satellite industry is clearly committed to revisit and revamp the role of satellite communications. The combination of satellite and terrestrial components to form a single/integrated telecom network has been regarded for a long time as a promising approach to significantly improve the delivery of communications services [1]. In many cases satellite networks complement the existing and the new terrestrial technologies by providing communication characteristics that cannot be supplied by other technologies. Satellite provides the means to support the expansion of the use cases towards other domains, especially global and highly reliable and secure networks [2]. What's more, satellite can be combined with a range of terrestrial technologies for the last mile communication such as WLAN, LTE or DSL, to create microsystems that can support multiple functions autonomously. In these cases, satellite network is only used for backhaul to the Internet, and the rest is provided by smart applications at the edge [2]. Integrated satellite-terrestrial solutions cost less in numbers of 5G use cases, taking costs and edge network deployments into consideration. The satellite network can provide the best and most comprehensive coverage for low-density populations, while the terrestrial network or the ground component can provide

© Springer Nature Singapore Pte Ltd. 2020
Q. Yu (Ed.): SINC 2019, CCIS 1169, pp. 56–65, 2020.
https://doi.org/10.1007/978-981-15-3442-3_6

the highest bandwidth and lowest cost coverage for high-density populations in urban environments [3]. Figure 1 illustrates some of the interesting satellite network use cases in 5G.

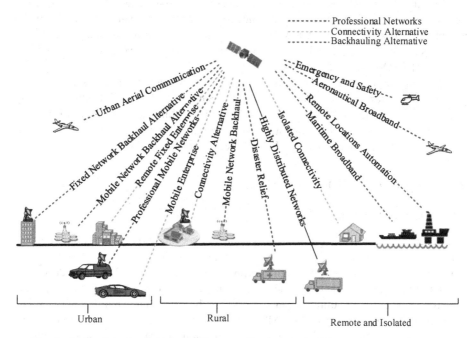

Fig. 1. Some of the interesting satellite network use cases in 5G (Source: Booz&Co) [2]

The increasing number of physical objects connected to the Internet at a remarkable rate brings the idea of the rapid evolution of the Internet of Things. It is one of the evolutionary directions of the Internet. Many of the objects that surround us will be on the network in one form or another. LEO satellites possesses irreplaceable functions in IoT applications. However, for the IoT vision to successfully emerge, the computing paradigm will need to go beyond traditional mobile computing scenarios that use smart phones and portables, and evolve into connecting everyday existing objects and embedding intelligence into our environment [4]. LEO-based IoT system, which is a realizable and powerful supplement to the terrestrial IoT network, has the advantages of low propagation delay, small propagation loss and global coverage.

Traditional architectures and network protocols for LEO-based IoT networks are not designed to support high level of scalability, high amount of traffic, efficiency and cost effective manner. To achieve such goals, emerging technologies such as SDN and NFV are being considered as technology enablers to provide adequate solutions. They bring flexibility which can be used to allow different objects connected to heterogeneous networks to communicate with each other, allowing simultaneous connections of various communication technologies. Network management decisions such as routing, scheduling can be done at the SDN controller and moreover, the programmability allows for any updates for new proposals.

Due to the high cost of satellite launches, it is currently hard to conduct research in real satellite networks. Software simulation approaches like OPNET or NS-3 are not capable of generating real data packets. Thus, SDN-based semi-physical emulation architecture, which is closer to the actual situation is constructed. The main aim of this design is to verify a satellite communication system can transmit user service with acceptable performance [5]. SDN structure separates the control and data layer, which promises it to be reconfigurable and scalable, providing an effective and helpful tool for research of LEO-based IoT networks.

The reminder of this paper is organized as follows. Section 2 introduces some background information including the combination of Satellite and terrestrial network, LEO-based IoT network and a brief review of SDN and NFV technologies. Section 3 illustrates the framework of a SDN-IoT emulation architecture. In Sect. 4, we discuss some use cases of this emulation architecture. Finally, Sect. 5 concludes this work.

2   Background and Related Works

2.1   LEO Constellation for IoT

IoT is a burgeoning paradigm that points out a novel direction of future internet, in which numerous heterogeneous networks containing different user data will be integrated transparently and seamlessly through appropriate protocol stacks [2, 6]. It is designed to support a massive number of devices, low data rate, low device power consumption, an extreme coverage and ultralow device cost [7]. There are a large number of potential necessities of constructing satellite IoT system such as to access in extreme topologies, to provide a cost-effective solution for IoT application in remote areas and to achieve global IoT service covering as a supplement and extension to the terrestrial IoT network. LEO-based IoT system is a realizable and powerful supplement to the terrestrial IoT networks. Generally, a LEO satellite constellation consists numbers of satellites in orbits of 500–2000 km [8]. LEO satellite constellations come into view to solve the covering problems of the terrestrial IoT network. They play a huge role in the following scenarios: IoT applications in extreme topographies, remote areas and where base stations are inaccessible.

Advances in satellite communications are being addressed from multiple angles. LEO satellite constellation technology has unique advantages. Due to the lower orbit altitude of LEO satellite constellation, it is more time efficient and more economical. In terms of propagation delay quantified by a round trip time (RTT), LEO satellite constellation has a RTT less than 100 ms [9]. What's more, the signal loss is supposed to be smaller thanks to the shorter distance of LEO satellite constellation. They possess more flexible payloads components to dynamically modify satellite antenna beam patterns in orbit to respond to market demand [10].

2.2   Overview of SDN and NFV

Software defined networking is a new networking paradigm that changes the limitations of current network infrastructures. It provides flexibility to network management by separating the network infrastructure into distinct planes [11]. It decouples the forwarding

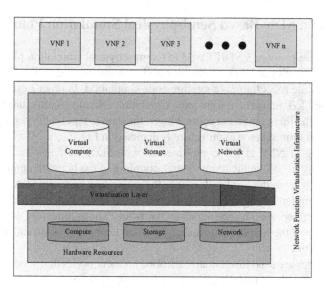

Fig. 2. NFV architecture

plane (data plane) and the network's control logic (control plane) traditionally coupled with one another [12]. The control plane is implemented in a logically centralized
controller (or network operating system), simplifying policy enforcement and network
configuration and evolution [13]. SDN is a new approach for network programmability where the network operator programs the controller to automatically manage data
plane devices and optimize network resource usage. This results in improved network
performance in terms of network management, control and data handling [14]. Network
function virtualization [15] is a network architecture concept of replacing dedicated
network appliances such as switches, routers, firewalls, to name a few, with software
running on commercial off-the-shelf servers [12]. It brings advantages in terms of energy
savings, load optimization and network scalability. Figure 2 shows a NFV architecture
which may consist of one or more virtual machines running different applications and
processes in network servers, switches, storage, or even cloud computing infrastructure,
instead of having custom hardware appliances for each network function [12]. NFV can
serve SDN by virtualizing the SDN controller to be rendered in the cloud thus allowing
dynamic migration of the controllers to the optimal locations while SDN can serve NFV
by providing programmable network connectivity between NFVs to achieve optimized
traffic engineering [16].

SDN and NFV solutions are expected to be relevant in addressing different challenges
in an IoT environment. Heterogeneous devices exchange data formats with diverse protocols. Lack of cooperation and capability mismatch between devices hinders the performance of network. SDN overcomes the heterogeneity of IoT devices so that the interoperability challenge in IoT can be solved. Another issue that needs to be addressed is the
discoverability in IoT devices. The ability to self-configure and adapt to the environment
without human intervention is a main factor of IoT devices, which is guaranteed by SDN
approaches. What's more, SDN can be used to solve problems of security maturity in
IoT applications.

3   Design of a SDN-Based Semi-physical Simulation Platform

In this section, we propose a SDN-based IoT semi-physical simulation platform which takes the characteristics of LEO satellite constellation into consideration. This platform supports large scale reconfigurable satellite node simulation. In order to simulate a specific LEO-based IoT scenario, a four-level simulation scheme architecture is designed, which contains the logic layer, the control layer, the data layer and the perception layer. The whole simulation scheme architecture is depicted in Fig. 3.

The enabling of NFV to IoT complimented with SDN increases the network efficiency and agility of IoT applications. The decoupling of the network control and management function from the hardware makes the simulation platform to be flexible and sustainable by implementing network functions in the cloud which reduces the dependency of emerging wireless technologies on hardware. The architecture we proposed in Fig. 3 is built upon the SDN architecture for IoT and considers a virtualized approach for the IoT network.

Physical nodes are used to simulate backbone nodes or nodes that need to be focused on in the network. Each node runs a client program that can communicate with the control system, receiving and executing the control information sent by the control system. Meanwhile, they feed back the simulation result information to the control system in real time. Various embedded devices and inter satellite links can also be added to the physical nodes to accommodate more types of physical nodes and real links for simulation.

The virtual system mainly consists of several servers, through which multiple virtual nodes are created. It can be expanded to any scale simulation model, and can be connected to the simulation network through virtual software switch. Therefore, it supports dynamic addition and deletion of virtual nodes, realizing the scalability and flexibility. Virtual nodes are mainly used to simulate the large-scale communication nodes in the scene and to achieve the purpose of large scale simulation.

3.1   Logic Layer

The logic layer includes a satellite backbone transmission network, an inter-satellite link, a satellite access network, a ground station and actual users. Based on the specific application scenario, a complete network model is established. The network model includes not only all network nodes, but also the specific connection plan, the link status corresponding to each moment, and the protocol stack.

Typical LEO-based IoT application scenarios are divided into two groups. (1) Delay-tolerant applications (DTAs). For instance, monitoring and forecasting applications. (2) Delay-sensitive applications (DSAs). For instance, enhanced supervisory control, data acquisition and military applications. The main network architecture includes TCP/IP protocol and Delay-tolerant networking (DTN) protocol architecture. The concept of DTA is a part of DTN, which is a novel communication structure to provide automated store-and-forward data communication services in networks [17]. One of the typical DTAs is water monitoring. DSAs are quite different scenarios that have stringent requirements (i.e., lower latency and higher reliability) from the DTAs. Smart grid and

Internet of Battle Things (IoBT) are representative application scenarios for civil and for military, respectively.

In logic layer, network architecture and each satellite node is determined. Satellite Tool Kit (STK) toolbox is used to design and model the simulation scene, including the antenna distribution, specific link conditions and constellation operation of each satellite. Aiming at the specific simulation object, model is established, including the network protocol used, the amount of data, data transmission scheme, etc. After the simulation scenario modeling is completed, the on-off information and link characteristic data of each link will be stored in the database, waiting for the controller to read.

3.2 Control Layer

The control layer mainly includes the control system and the SDN controller. The control layer receives the input of the description of network structure from the logic layer, and drives the data flow simulation in data layer. STK provides analysis engine to analyze specific scenarios and generates necessary position and attitude information about satellites. It also provides visualization data, which plays a certain role in elaborate analysis of routing scenarios.

In the LEO-based IoT network, the topological relationship between nodes changes frequently and complexly. An accurate simulation of the LEO satellite constellation topology is a key element of the emulation architecture. Thus, the SDN controller is introduced for precise node topology control. In the area of scalability, the effectiveness of the proposed architecture is highlighted. The separation of the network functions coupled with SDN will allow resources to be automatically allocated by the cloud infrastructure to handle increase load. It is responsible for the control of the software definition switch in the simulation process. The simulation scene topology information file is generated by the STK and imported to the controller. The controller sends the flow table to the OpenvSwitch. All users are connected to the switch, which continuously changes the on-off relationship between the network ports according to the flow table sent by the controller in real time to simulate the topology changes in the scenario.

3.3 Data Layer

LEO constellation involves a large number of satellites. Traditional virtual machine takes up too much resource when making such a large scale simulation. For this reason, we choose docker to virtualize nodes in the system. In the simulation platform, each satellite node runs the same routing and forwarding program. After compiling the forwarding program and adding it to the basic ubuntu image, it realizes the large-scale flexible deployment of virtual nodes to simulate LEO satellite.

The terrestrial IoT communication system is based on the USRP and LabView Platform. Terrestrial IoT network comprises of various kinds of physical devices like sensors, RFID, smart meters, WSN, NFC etc. that allows communication with the IoT Gateway. They sense and collect data from various kinds of devices in an intelligent way.

The system is mainly composed of the NI USRP hardware system (or NI's satellite monitoring and control data transmission integrated baseband unit) and the LabVIEW programming environment running on the computer. The two are connected through Gigabit Ethernet.

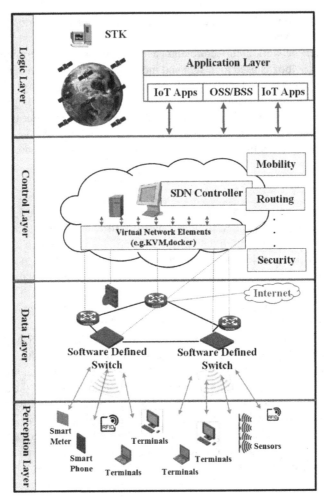

Fig. 3. The whole semi-physical simulation platform. The platform is divided into multi-layers to realize the separation of network control and data transmission.

4   Case Study

4.1   Simulation of Satellite Routing Algorithms

The system is capable of simulating satellite routing algorithms. In the whole routing scenario, the topology and the link characteristics are continuously changing. In the data layer, IoT nodes collect information and send them regularly to the gateway. Gateway send this information to the satellite above it. After transferring between the satellites, it comes to the IoT data center. LEO satellites convergence through DRA/OSPF/CGR algorithms, determine multi-hop routing and eventually send data packets back to the ground station for reception.

In this case, this paper selects the $6 \times 12/6/0°$ polar orbit satellite constellation as the model for the simulation platform. This model is divided into 6 orbits; each orbit has 12 satellites. The angular difference between adjacent two orbits is 30°. For the sake of simplicity and the cross-slit link, the angle of between the orbits and the equatorial plane is 90°. An inter-satellite link is established between each satellite and four adjacent satellites. The whole system is shown in Fig. 4.

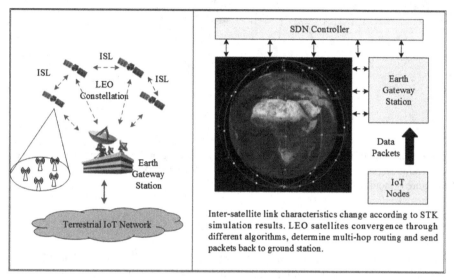

Fig. 4. Simulation of different routing algorithms

4.2   Simulation of Combined LEO-Terrestrial Mobile Network

SDN concepts have been adopted in mobile network architectures in several ways. Figure 5 depicts an illustrative view of a SDN-based mobile network [18], which promotes a SDN/NFV-enabled satellite ground segment system for realizing an End-to-End traffic engineering (TE) use case. The mobile core network control functions together with specific TE functions for the transport network are realized as applications running on top of a SDN controller.

The system demonstrated in this paper supports the simulation of the LEO-terrestrial mobile network embracing SDN technologies proposed in [15]. Mobile core network control functions are realized as applications running on top of a SDN controller, which is responsible for managing the network elements that provide the packet switching and forwarding capabilities. In particular, the openflow switch abstraction model is considered to model the operation of the virtual satellite network as seen from an external controller entity. The virtual satellite network comprises LEO satellites and gateways, which are interconnected with the outside world. Mobile terminals are linked to RAN nodes.

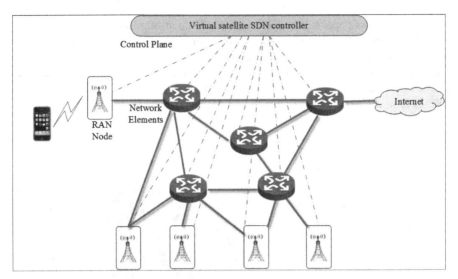

Fig. 5. Simulation of a SDN/NFV-enabled satellite ground segment system for realizing a end-to-end traffic engineering in a combined terrestrial-satellite network use for mobile backhauling. This figure is modified from [18].

5 Conclusion and Future Work

It is essential to build a mature and complete LEO-based IoT network. The adoption of SDN and NFV technologies into the satellite domain is seen as a key facilitator to enhance the delivery of satellite communications services and achieve a better integration of the satellite component within the 5G ecosystem. In this paper, an SDN-based LEO-terrestrial IoT emulation architecture is proposed. It gives the overall architecture design and physical model. The emulation architecture has the characteristics of large-scale, multi-level, flexible, reconfigurable and maintainable. It provides credible and reliable experimental tools for the development of LEO-based IoT network. With booming development in IoT environment and satellite communications, potential use of constellations of satellites for IoT applications is of growing interest. Constructing a SDN-based emulation architecture is not only conductive to verifying the innovation of IoT technology, but also conductive to provide reference for future development and making this topic become a reliable cost-benefit solution.

References

1. ETSI, TS: 103 124: Satellite earth stations and systems (SES) combined satellite and terrestrial networks scenarios. ETSI, France (2013)
2. Corici, M., et al.: Assessing satellite-terrestrial integration opportunities in the 5G environment (2016). https://artes.esa.int/sites/default/files/Whitepaper
3. Kim, S., Kim, H.W., Kang, K., Ahn, D.S.: Performance enhancement in future mobile satellite broadcasting services. IEEE Commun. Mag. **46**, 118–124 (2008)

4. Gubbi, J., Buyya, R., Marusic, S., et al.: Internet of Things (IoT): a vision, architectural elements, and future directions. Future Gener. Comput. Syst. **29**(7), 1645–1660 (2013)
5. Gao, S., Liu, Y., Fan, L., Cao, Y., Long, G.: Model-based semi-physical simulation platform architecting for satellite communication system. In: 2018 13th Annual Conference on System of Systems Engineering (SoSE), Paris, pp. 379–386 (2018). https://doi.org/10.1109/sysose.2018.8428761
6. Khan, F.: Mobile internet from the heavens. CoRR, vol. abs/1508.02383 (2015)
7. Wan, L., Zhang, Z., Wang, J.: Demonstrability of narrowband internet of things technology in advanced metering infrastructure. J. Wirel. Commun. Netw. **2019**, 2 (2019). https://doi.org/10.1186/s13638-018-1323-y
8. Qu, Z., Zhang, G., Xie, J.: LEO satellite constellation for Internet of Things. IEEE Access **5**, 18391–18401 (2017)
9. Sanctis, M.D., Cianca, E., Bisio, I., et al.: Satellite communications supporting internet of remote things. IEEE Internet Things J. **3**(1), 113–123 (2015)
10. ARTES programme: Flexible payloads key for satcom industry, July 2015. https://artes.esa.int/news/flexible-payloads-key-satcom-industry
11. Wickboldt, J.A., Paim, W., Isolani, P.H., et al.: Software-defined networking: management requirements and challenges. IEEE Commun. Mag. **53**(1), 278–285 (2015)
12. Ojo, M., Adami, D., Giordano, S.: A SDN-IoT architecture with NFV implementation. In: 2016 IEEE Globecom Workshops (GC Wkshps), Washington, DC, pp. 1–6 (2016)
13. Kreutz, D., Ramos, F.M.V., Verissimo, P., et al.: Software-defined networking: a comprehensive survey. Proc. IEEE **103**(1), 10–13 (2014)
14. Kim, H., Feamster, N.: Improving network management with software defined networking. IEEE Commun. Mag. **51**(2), 114–119 (2013)
15. Han, B., Gopalakrishnan, V., Ji, L., et al.: Network function virtualization: challenges and opportunities for innovations. IEEE Commun. Mag. **53**(2), 90–97 (2015)
16. Li, Y., Chen, M.: Software-defined network function virtualization: a survey. IEEE Access **3**, 2542–2553 (2015)
17. Bedon, H., Miguel, C., Fernandez, A., Park, J.S.: A DTN system for nanosatellite-based sensor networks using a new ALOHA multiple access with gateway priority. Smart CR **3**, 383–396 (2013)
18. Ferrus, R., Sallent, O., Ahmed, T., Fedrizzi, R.: Towards SDN/NFV-enabled satellite ground segment systems: end-to-end traffic engineering use case. In: 2017 IEEE International Conference on Communications Workshops (ICC Workshops), Paris, pp. 888–893 (2017)

A Link-Estimation Based Multi-CDSs Scheduling Mechanism for FANET Topology Maintenance

Xiaohan Qi[1], Xinyi Gu[1], Qinyu Zhang[1,2], and Zhihua Yang[1,2(✉)]

[1] Communication Engineering Research Center, Harbin Institute of Technology
Shenzhen Graduate School, Shenzhen, Guangdong, China
{qixiaohan,guxinyi}@stu.hit.edu.cn, {zqy,yangzhihua}@hit.edu.cn
[2] Pengcheng Laboratory, Shenzhen, Guangdong, China

Abstract. A Connected Dominating Set (CDS) is a useful method for degrading major routing and forwarding operations in the network, which is widely applied in the mobile ad hoc networks. In a flying ad hoc network (FANET), however, high dynamics of nodes produce considerably large challenges on the topology maintenance due to rapidly time-varying connections between nodes, which will lead to a huge computation latency and overheads if exploiting current CDS algorithms. In this paper, therefore, we proposed a connection estimation-based topology control mechanism to achieve efficient maintenance of connectivity in the network. In particular, the proposed algorithm could provide a stable and effective virtual backbone sub-net in a fast changing topology of FANET, by flexibly scheduling multiple Minimum Connected Dominating Sets (MCDS) with a very efficient method. The simulation results show that, compared with typical single CDS method, the proposed algorithm presents better performances in obviously dynamic environments with respect to updating counts and rate of successful updates.

Keywords: Connected Dominating Set · FANET · Connection estimation · Topology control · Backbone

1 Introduction

In these years, a flying ad hoc network (FANET) [1,2] is an emerging variety of mobile self-organizing dynamic networks with highly autonomy, which is composed of multiple unmanned aerial vehicle (UAV) nodes in the absence of infrastructure or central control entity. In the network, each node plays the roles of a host and a router at the same time. Due to its advantages of self-organizing, self-healing and fast deployments, FANET is especially suitable for resisting harsh environments in military aviation scenarios [3]. In the network, two nodes can communicate directly with each other within a constrained communication range. Once the distance beyond the communication range, source nodes require to reach the destination nodes through a series of intermediate

© Springer Nature Singapore Pte Ltd. 2020
Q. Yu (Ed.): SINC 2019, CCIS 1169, pp. 66–86, 2020.
https://doi.org/10.1007/978-981-15-3442-3_7

nodes. Moreover, in a certain complex scenarios, such as obstacles avoidance and low-altitude surveillance, quite a lot of types of clutters result in frequently intermittent connectivity between nodes, which will result in huge challenges on the topology control of network. At present, building a backbone network is an effective method to solve such problems. Typically, in a flying ad hoc network, nodes within a backbone network will take the task of maintaining the entire network with a certain constrain of limited time, which is fundamentally different with terrestrial ad hoc networks. Therefore, those backbone nodes are time-changing and thus the backbone network in FANET is considered as a virtual backbone network (VBN).

Currently, constructing a connected dominating set (CDS) is widely accepted for an efficient mechanism to build a virtual backbone network. In graph theory, a dominating set (DS) for a graph $G = (V, E)$ is a subset D of V such that every vertex not in D is adjacent to at least one member of D. A DS is a set of nodes constituting a backbone network of a FANET. All nodes in the network are either exactly the dominating nodes or connected with at least one dominating node. Generally, the nodes in the dominating set are called dominators, while others are called dominatees. If the dominators are connected, the DS is called a CDS. The CDS provides a simple and straightforward way to construct a VBN, which could be evaluated through a group of quantitative metrics. In particular, only a part of routing information needs to be maintained for the dominators of CDS. Furthermore, by using one CDS based VBN, recalculating the routing table is not essential for the network as long as the topology is changed. Generally, a smaller size of VBN can simplify the routing procedure and reduce routing cost for all the nodes. Therefore, a CDS with small number of dominators is expected in the construction, which could reduce computation and communication overheads. Currently, there are lots of existing works on the minimum connected dominating set (MCDS), such as the methods of mathematical modeling [4], maximum independent set [5], minimum spanning tree [6], pruning and other algorithms, respectively. In particular, the proposed method of MCDS in [7] is widely used for effectively constructing a virtual backbone network in the mobile ad hoc networks.

In a FANET, however, highly relative motions between nodes in fast changing scenarios produce considerably large challenges on the construction and maintenance of a virtual backbone network due to rapidly time-varying property of topology. For example, a pre-fixed connected dominating set could not ensure permanent connectivity of VBN since partial nodes in the CDS may be failed due to exhausted energy or disrupted link. Once the on-duty CDS is disjoined due to failures of partial dominator nodes, a candidate group of nodes should be rapidly supplemented into the CDS, even a completely new CDS requires to be substituted for the current CDS in extremely severe conditions. Most existing topology maintenance algorithms mainly focus on topology reconstruction, which will lead to a relatively large computation overheads and latency.

In this paper, we propose a multi-MCDS based topology control algorithm for a flying ad hoc network, which could effectively maintain network connectivity by constructing a VBN with a multi-MCDS scheduling mechanism. It is the first work of topology control on the flying ad hoc network we have found at present. In particular, this paper has the following contributions:

(a) we built a swarming-oriented mobility model of UAV node in FANET, which could be suitable for describing the intelligent behavior of a clustered UAV team. Furthermore, we proposed a estimation mechanism of current topology status by estimating the connectivity duration between nodes with node's flying velocity.

(b) we designed a time-varying VBN by periodically scheduling multiple CDSs calculated in prior according to the current status of topology. Compared with a single CDS method, the proposed method could maintain more stable connectivity in dynamic scenarios with obviously reduced overheads in term of the amounts of replaced nodes.

The remainder of the paper is organized as follow. Section 2 overviews the related work. Section 3 describes the mobility and system model. In Sect. 4, the problem formulation is presented and the proposed algorithm is proposed in Sect. 5. Section 6 presents the simulation results. Section 7 concludes the paper finally.

2   Related Works

Generally, the topology maintenance is considered as the detection of network connectivity and the maintenance of stability of virtual backbone network. Typically, the maintenance of the topology is a periodical procedure, which is triggered by different criterion in a single cycle duration [8]. The main purpose of topology maintenance is to guarantee the VBN to sustain the stable communication of network, by dealing with the dynamic changes of the network and extending the life cycle of the network [9–11]. To solve this problem, several works maintain the network by constructing a fault tolerant K-connected and M-dominating set, which can tolerate failures of multiple nodes. In the algorithm, there are k disjoint paths between any two nodes in the graph, thus any removal of k-1 nodes will not affect the connectivity of the derived sub-graph. Dai et al. earlier proposed the construction of multi-connected and multi-dominating set in mobile ad-hoc networks. Three different algorithms were proposed in [14], but these algorithms mainly address the special case of k equaling to m. In addition, the proposed algorithms depend closely on the parameters such as network size and node density, which are difficult to obtain in practical applications. In recent years, building (k, m)-CDS develops into one of major protocols to maintain the network, which attracted more works. Nevertheless, more active nodes will be generated leading to great network overheads, which recurs the deteriorated performance of the maintenance algorithm over time. A community-based greedy algorithm is applied for searching the whole network information

[12]. The proposed algorithm selects nodes with better performance to construct *(k, m)*-CDS, and deletes the redundant nodes to minimize the number of dominating nodes for a smaller scale. As a result, a better set scale is achieved, while the time and message complexity are higher in the dense network.

In [13], Liao et al. present a topology maintenance algorithm based on clustering, which is triggered by the time. When the existing cluster heads run after a period of time, the algorithm will be regularly triggered in order to balance the energy consumption of each node. Nevertheless, the algorithm cannot autonomously responds to the fault due to its time-driven mechanism. The algorithm in [15] constructed the backbone network through constructing k-hop CDS. Intuitively, the larger number of k, the smaller size of CDS. When k is infinite, there will be only one dominating node in the network. During the maintenance phase, the algorithm maintains the network by detecting the change of network and revising the value of k. A large value of k will cause a part of dominators to carry heavier forwarding messages, which will easily lead to network segmentations.

3   System Model

3.1   Mobility Model

Typically, a flying ad hoc network can be regarded as a special kind of mobile ad hoc network (MANET) consisted of a group of flying nodes with relatively high mobility. Currently, a series of motion models reflecting the node's moving characteristics, are currently used in MANET, especially in terrestrial scenarios, i.e., random mobility model, time-limited model, space constrained model and environment constrained model. However, these models have incapability of formulating the swarming properties of UAV nodes in a FANET, especially for certain types of missions of intelligent multi-UAV team. Therefore, we consider the multi-UAV mobility in a FANET as a type of cluster movements, in which each node cooperates with each other during the movement to accomplish the task. In actual motions, nodes tend to move toward a common direction, and adjust their directions and movements according to the states of neighbor nodes. In this paper, we proposed a *boid*-based [10] mobility model, called as SYN-*boid*, to describe the motional property of FANET nodes, which presents a typical behavior of a UAV swarming team. Here, we give several related assumptions. Firstly, we assume that all of nodes have the same physical characteristics and same detection ranges. Besides, we consider that within a constrained communication range two nodes can communicate with each other directly.

In the proposed SYN-*boid* model, four defined rules, i.e., Cohesion, Separation, Alignment and Synchronization, are exploited for constraining the motions of nodes in FANET. Compared with general *boid* model, in the SYN-*boid* model, we introduced a novel rule, called as synchronization rule, in order to avoid flight confusions with extra adjustment costs. During the flight, each node will make motion by exchanging a certain quantity of information with each other under the four rules in SYN-*boid* model. Each rule generates a corresponding motion

(a) Initial Stage   (b) Intermediate Stage   (c) Stabilizing Stage

Fig. 1. Swarm mobility state.

component, and each component has a different weight depending on the scene and mission requirement. As a result, all of the components decide the motion of node. The motion state of the node is shown in Fig. 1. In the beginning, nodes are placed in the area randomly. During flying, the motion state of the node cluster gradually reaches synchronization.

$$\overrightarrow{M_{n+1}} = \alpha\overrightarrow{M_n^1} + \beta\overrightarrow{M_n^2} + \chi\overrightarrow{M_n^3} + \delta\overrightarrow{M_n^4} + \varepsilon\overrightarrow{M_n^5} \tag{1}$$

in which $\overrightarrow{M_n^k}$ represents different component determined by five mobility rules at current moment. And $\alpha, \beta, \chi, \delta, \varepsilon$ are different weighting factors with constrain of $\alpha + \beta + \chi + \delta + \varepsilon = 1$. The detailed descriptions of five rules are shown as follows.

Inertia Rule. In a realistic scene, the motion status at current moment is also one of the determining conditions for next moment. Therefore, the motion state at current time is taken as the inertial movement component of the next moment, as

$$\overrightarrow{M_n^1} = \overrightarrow{M_n} \tag{2}$$

Cohesion Rule. Cohesion rule attempts to let nodes gather with nearby flock-mate modes, so that the swarming could moves toward an average position.

$$\overrightarrow{M_n^{'2}} = \sum_j \overrightarrow{CC_j'} \Big/ N \tag{3}$$

and

$$\overrightarrow{M_n^2} = \overrightarrow{M_n^{'2}} \Big/ \left|\overrightarrow{M_n^{'2}}\right| \tag{4}$$

where C is the position of node i at current moment, and C_j' is the position of $i's$ neighboring node j, respectively.

Separation Rule. The separation rule attempts to make local node avoiding collisions with nearby flock-mates, as

$$\overrightarrow{M_n^{'3}} = \sum_j \overrightarrow{C_k'C} \Big/ N \tag{5}$$

and

$$\overrightarrow{M_n^3} = \left| \overrightarrow{M_n'^3} \right| \tag{6}$$

where $\overrightarrow{C_k^i}$ represents direction of the nodes in the repulsion region of node i.

Alignment Rule. Alignment rule is designed for keeping nodes match the average velocity with nearby flock-mates, as

$$\overrightarrow{M_n^4} = \sum_g \overrightarrow{C_g^i} \bigg/ N \tag{7}$$

and

$$\overrightarrow{M_n^4} = \left| \overrightarrow{M_n'^4} \right| \tag{8}$$

where $\overrightarrow{C_k^i}$ is the average direction of nodes in $i's$ effective sensing range. Is the direction of $i's$ detectable nodes g.

Synchronization Rule. A parameter syn_i is defined into the velocity calculation to describe the synchronization degree of the global swarming.

$$\overrightarrow{M_n^4} = \overrightarrow{M_n} syn_i \tag{9}$$

$$V(n+1) = V_{max} e^{\beta[syn_i(n)-1]} \tag{10}$$

where syn_i represents the synchronization coefficient of $i's$ nearby nodes. In particular, a larger value of syn_i means a better synchronization. When all the neighbor nodes in a certain node's communication radius move in the same motion, the value of syn_i equals to 1.

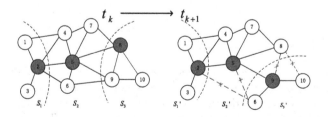

Fig. 2. Updates of CDS in a time-varying topology.

3.2 Network Model

In this paper, we define a topology of FANET as an un-directional graph $G(V^t, E^t)$. In the graph, V^t represents the set of flight nodes at time t, and $V^t = \{v_1, v_2, v_3, \ldots, v_n\}$. On the other side, E^t represents the connections between two nodes, denoted by $E^t = \{e_1, e_2, e_3 \ldots e_n\}$. Typically, a connected dominating set is a sub-graph of V^t. In this paper, we will construct multiple CDSs for maintaining the virtual backbone network, which needs to be updated with the changing topology. For example, we describe the update process of CDS at adjacent time slots in Fig. 2.

According to the definitions of CDS, we have two following observations.

- There are a certain number of minimum dominating sets in a connected graph.
- If a failed node $G(V, E)$ and the links $\{e = v_i^t v_j^t | v_i^t, v_j^t \in V_{cds}^t\}$ are effective, the current CDS is still effective.

Theorem 1. Both CDS and MCDS. Problems are NP-complete problems.

Proof: We transform the dominating set problem to the vertex-cover problem. If DS problem is a NP problem, then MCDS is a NP problem. *Vertex-Cover Problem:* if $G(V, E)$ is a unidirectional graph, we denote that a subset D of V is a vertex-cover, which is meant that one of the endpoints in E belong to D. Given a graph G and a budget b, we attempt to find a vertex-cover whose size is not larger than b. For each edge $e_{u,v}$, we add an auxiliary vertex w and two another edges $e_{u,w}$ and $e_{v,w}$ to get a new graph G_0. Then, the vertex-cover problem in G can be transferred as dominating set problem in G_0.

Corollary 1. The minimum vertex-cover of G corresponds to the minimum dominating set of G_0.

Proof: since S is adjacent to all the edges in E, S contains or is adjacent to all of the nodes in G. The newly added node w is derived from the edge $e_{u,v}$, so w is also adjacent to the point in S. It is proved that S constitutes a dominating set of G_0.

Corollary 2. The minimum dominant set S of G_0 corresponds to the minimum vertex-cover of G.

Proof: For each new w in G_0, if w belongs to S', its adjacent nodes must not belong to S at the same time, otherwise S is not the minimum. Therefore, we can replace w with the nodes in $Neilist(w)$ that does not belong to S. It is obvious the new set S' is still a minimum dominating set. Thererfore, we only consider that all points belong to G.

For the obtained S, all newly added nodes are adjacent to the nodes in S. Therefore, all edges $e_{u,v}$ in G are adjacent to the nodes in S. As a result, the minimum dominating set S is a minimum vertex-cover of G. If the dominant set (G_0, b) has no solution, the point coverage (G, b) has no solution. Therefore, if the dominating set problem is not NP-complete, the point covering problem is not an NP-complete one. It is known that vertex-cover problem is NP-complete, thus the dominating set problem is also NP-complete.

Since the MCDS problem is NP-hard, we adopted a distributed approximately optimized algorithm based on the minimum spanning tree algorithm, instead of a centralized method. The applied symbols in the algorithm are shown in Table 1. The initialization phase is divided into three parts, including neighbor information exchange, node competition process and node data structure update respectively. Each node sends the periodic messages with the maximum power. When receiving the message, the neighboring nodes respectively calculate their own weights and continue to broadcast their own weights. Until all nodes in the network finish the process. At this moment, each node contains the weight and ID information of the neighbor nodes. During the construction phase of the virtual backbone network, the algorithm is executed according to the information of the neighbor nodes obtained in the previous step to determine the status of each node. The network needs to detect changes in nodes and CDS at adjacent time slot according to the data structure as shown in Fig. 3 in order to schedule and maintain the CDSs in a distributed way.

- *Neighbor list:* it is denoted by $Nlist(v_i)$, including ID and the indicator whether backbone nodes in the $v_i's$ neighbors. which are respectively denoted by $Nlist(v_i).ID$ and $Nlist(v_i).flag$.
- *Neighbor dominator list*: it is denoted by $Domlist(v_i)$, and refers to the dominator list in $Nlist(v_i)$.
- *Parent field of domintee:* represents the dominator which dominates v, donated by $v.parent$, v is domintee.

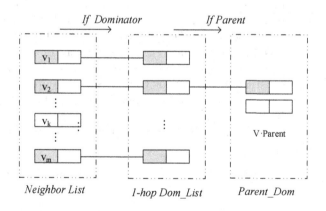

Fig. 3. Node data structure.

Table 1. Symbols description.

Notation	Representation of the symbol or symbol
S_{cds}	The set of CDSs in the graph
$ESm^{t_c}(S_{cds})$	CDS with maximum energy at time t_c
$Em^{t_c}(cds_i)$	Energy of CDS_i
V_{cds_k}	Nodes in CDS_k
E_{cds_k}	Links of nodes in CDS_k
$N_1(v)$	Node lists in $v's$ 1-hop range
$N_2(v)$	Node lists in $v's$ 2-hop range
$v.parent$	Dominator of v
$Nlist(v)$	Neighbor list of v
$Domlist(v)$	Dominator nodes in $N_1(v)$
CDS_{ct}	The CDS of graph at current time
$T_{\_\cos t}$	Total update time overhead
$P_{\_total}$	Total successful update probability
r_{t_i}	Number of failure nodes at ith time slot
$r_{\_FN_i}$	Number of dominatees need to update
$r_{\_NC_i}$	Number of dominators need to update
LT_k^i	Estimated connected time of k and i

4   Problem Formulation

Given a directed graph $G^t(V^t, E^t)$, we attempt to build a backbone network with less nodes and maintain it with less update times. In the construction phase, we adapt a heuristic MST-CDS [16] algorithm to build the backbone network, which has been proved to have the shortest path. In the maintenance phase, we should detect the changes of the graph at every time interval τ. When a topology change is detected, the current CDS needs to be updated. We hope that the update will be completed in a short time. Here, we define a total update time cost T_{cost} as the total numbers of updates over a successive of snapshots. In the actual implementation, we hope to get smaller T_{cost} and higher update success probability.

Total Update Overhead: the number of updated nodes during the whole lifetime of the network.

$$C_{\_node}(v_i) = \begin{cases} 0 & v_i \ will \ be \ updated \\ 1 & v_i \ won't \ be \ updated \end{cases} \tag{11}$$

where

$$T_{\_\cos t} = \sum_{t=1}^{i} \sum_{j=1}^{n_{BN}} \sum_{k=1}^{N-n_{BN}} C_{\_node}(v_{k,j}) \tag{12}$$

in which

$$1 \le i \le T, 1 \le j, k \le N$$

and t is the number of time slots sampled with interval time τ. In particular, n_{BN} signifies the number of backbone nodes during a certain time slot. Here, $C_{\_node}(v_{k,j})$ is a flag on whether node $v_{k,j}$ needs to be updated. N is the total number of nodes in the graph. It is noted that, $T_{\_\cos t}$ is one important metric of the algorithm performance, namely, the total update cost. The network maintenance process is triggered by the events of node failure or the expired time period. It is realized by scheduling the status of each node, such as dominator and dominatee. In fact, the high dynamic of the node may cause the failure of the maintenance process. The time sequence of maintenance implementation is shown in Fig. 4. In specific, we assume that the following three conditions will trigger the maintenance algorithm.

- The edges in E_{cds} or nodes in V_{cds} change.
- A node quits from the network.
- A new node joins the network.

We consider that the successful update rate of node is an indicator to evaluate the performance of the algorithm. In order to describe the effectiveness of the maintenance algorithm, we define the total successful update rate as the ratio of successfully updated nodes to total failed nodes.

Total Successful Update Probability: the updated probability of failure nodes in the network.

$$P_{\_total} = \frac{\sum_{i=1}^{k} h_{t_i}}{\sum_{i=1}^{k} r_{t_i}} = \frac{\sum_{i=1}^{k} h_{t_i}}{\sum_{i=1}^{k} r_{\_NC_i} + \sum_{i=1}^{k} r_{\_FN_i}} \tag{13}$$

where h_{t_i} represents the number of successful restoring nodes, and r_{t_i} is the number of nodes that make the current VBN unavailable, $r_{\_NC_i}$ and $r_{\_FN_i}$ are faulted dominators and domintees respectively.

5   Algorithm Description

5.1   Connection Estimation Algorithm

In order to reduce the renewal time of the backbone network, we use the parameter LT to describe the connection time of E_{cds_k}. The parameter of LT is estimated from current status of two nodes, which includes velocity, location and direction of movements, as shown in Fig. 5. When judging the connected time of two nodes, we consider one of these nodes as the center, then the relative status of i and neighbor node k is detected with time interval t_p. Because the node is making a continuous movement according to the SYN-*boid* rules, the state of the current moment determines the next movements. Therefore, by comparing the relative motion of adjacent time slots, the connectivity at the next moment can be estimated. Typically, the link time of two nodes LT_k^i can be deduced as

$$LT_k^{i\,2} = \frac{R_i^{\,2} - \vec{l}_m^{\,2}}{\left|\vec{v}_m^i\right|^2 + \left|\vec{v}_m^k\right|^2 - 2\left|\vec{v}_m^i\right|\left|\vec{v}_m^k\right|\cos(\theta_m^i - \theta_m^k)} \tag{14}$$

in which $\vec{v}_m^i$ and $\vec{v}_m^k$ represent the velocity of node i and k at time m, respectively. In particular, θ_m^i and θ_m^i represent the actual direction of movement of node i and k respectively. In specific,

$$|\vec{l}_m| = \sqrt{|\vec{l}_{m-1}| + |\vec{f}_m| + 2|\vec{l}_{m-1}||\vec{f}_m|\cos\varphi_m} \tag{15}$$

and

$$\varphi_m = \arccos\frac{|\vec{l}_{m-1}|^2 + |\vec{f}_m|^2 - |\vec{l}_m|^2}{2|\vec{l}_{m-1}||\vec{f}_m|} \tag{16}$$

Moreover,

$$\left(\left|\vec{v}_m^i\right|^2 + \left|\vec{v}_m^k\right|^2 - 2\left|\vec{v}_m^i\right|\left|\vec{v}_m^k\right|\cos(\theta_m^i - \theta_m^k)\right)LT_k^{i\,2} + \vec{l}_m^{\,2} = R_i^{\,2} \tag{17}$$

Here, $\vec{f}_m$ is the relative velocity vector of nodes, $\vec{l}_m$ is the relative distance vector of two nodes at time m, and φ_m is the angle between two vectors at time m, respectively.

The obtained estimated connectivity time LT_k^i is used as a trigger for maintaining the algorithm in the process to maintain the network. In FANET, nodes in the network may fail due to environmental factors or high dynamic of movements. When the sub-graph deduced by virtual backbone network is not connected, the maintenance algorithm need to work to maintain the stability of whole network. In particular, the node periodically broadcasts its own state information to obtain the state of the neighboring node and detect whether the current CDS needs to be maintained.

5.2   Multi-MCDS Construction Phase

In a FANET network, finding those smaller connected dominating sets will be expected with a corresponding maintenance algorithms in order to maintain network connectivity. However, a general maintenance algorithms could only detect the changes of nodes, without the required capability of rapid response to the topological changes of network. In this section, therefore, we propose a backbone construction algorithm based on multiple minimum connected dominating sets, which could provide a temporal plan for scheduling these connected dominating sets to reduce the latency waiting for connectivity.

Definition 1. (Dominating Set) Given graph $G=(\mathrm{V}, \mathrm{E})$, $v' \in V$ is a DS of G, only if $\forall(u, v) \in E$, either $u \in v'$ or $v \in v'$ is true.

Definition 2. (Connected Dominating Set) is a CDS of G if (1) S is a DS and (2) a graph induced by S is connected.

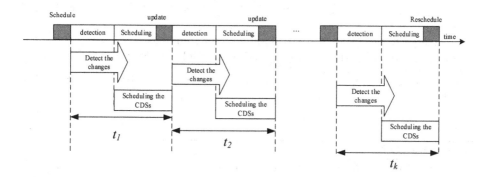

Fig. 4. Connected time estimation process.

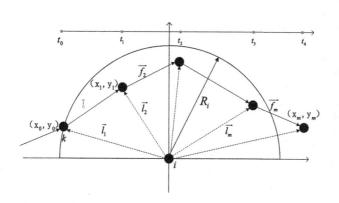

Fig. 5. Process of maintaining the network.

Definition 3. $D(u)$ for any node u in graph $G = (V, E)$, the number of $u's$ neighbors nodes is the degree of u, referred to as $D(u)$. In this paper, we design a distributed minimum spanning tree based algorithm to traverse the connected dominating set, as shown in Algorithm 1. In particular, the algorithm will obtain a minimum dominating set, which could be used for finding the minimum weight of backbone nodes. For example, we chose 40 random nodes to generate a MCDS, which is shown in Fig. 6. In the initial stage, each node broadcasts its own information with maximum power, and determines node degree and $N_1(v)$ according to the acquired information of neighboring nodes. During the construction of MST-CDS, the spanning tree is constructed step by step by selecting nodes with larger degree as follows.

- For a connected graph $G = (V, E)$, each node in V periodically broadcasts its own information and assigns a value to each edge belonging to E with $W(e_{u,v}) = D(u) + D(v)$.
- Triggered from any vertex, the minimum spanning tree algorithm is used to solve the spanning tree with the maximum weight $W(S) = \max_s \sum_{e \in S} W(e)$.
- By removing the nodes with degree of 1, the remaining nodes constitute the CDS.

Algorithm 1. MST-CDS

Require: The set of MCDSs at t-th time slot, S_{cds}^t
Ensure: Topology of current network.
1: Initial Calculate all edge weights in the graph
2: **for** $x = 1 : N$ **do**
3:    $v = v_{st}^x$
4:    $U = f_{\max}[w^t(N_1^t(v))]$ $//\{v \in V_{NB}, u \notin V_{NB} | u \in U\}$
5:    **if** $number(u) = 0$ //the only node is u_1 **then**
6:        $vector(i) = (B(i) - A(i))/norm(B(i) - A(i))$
7:    **else**
8:        $u = \arg\max(E^t(u_k)),\ k = 1 : size(U)$ $u_k \in U$
9:    **end if**
10:   **if** $V_{cds}^x = V$ **then**
11:       $V_{cds}^x \leftarrow u; E_{cds}^x \leftarrow <u, v>$
12:   **end if**
13:   **if** $D(u) = D(v) = 1$ and $e_{u,v}$ exist **then**
14:       Delete u and v
15:   **end if**
16: **end for**
17: **if** $V_{cds}^m == V_{cds}^n$ **then**
18:    Deleted V_{cds}^n
19:    Refresh V_{cds}^x, E_{cds}^x
20: **end if**
21: Return S_{cds}

5.3 Multi-MCDSs Scheduling Phase

In this phase, we find a group of MCDSs based on MST-CDS algorithm, which is denoted by $S_{cds} = \{CDS_1, CDS_2, ...CDS_k\}$ at t_c time slot. In order to reduce the reconfiguration time of backbone network, we propose a scheduling algorithm or coordinating Multiple MCDS in duty. Let $Em^{t_c}(CDS_i) = E^{t_c}(v)$ be the minimum energy of CDS_i, in which v is the node with minimum energy in CDS_i. $ESm^{t_c}(S_{cds})$ represents the maximum energy of S_{cds}, and

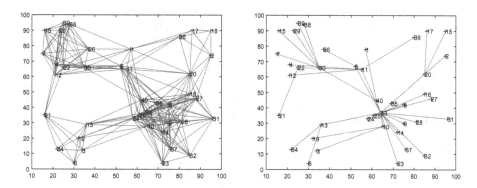

Fig. 6. Construction of MCDS. (a). connected graph. (b). corresponding backbone subset.

Algorithm 2. Multi-MCDS Scheduling algorithm

Require: S_{cds}, t_c, $G^t(V^t, E^t)$
Ensure: Scheduling of S_{cds}
1: Initialize $CDS_{t_c} \leftarrow ESm^{t_c}(S_{cds})$
2: **if** $number(V) > 1$ **then**
3:     Update $E_{CDS}, V_{CDS}, S_{cds}$
4:     **if** E_{CDS}, V_{CDS} change at time t_c **then**
5:         **if** $v_i \in V_{CDS_{ct}}$ fails **then**
6:             $\{CDS_k\} = \arg\{v_i \in CDS_k\}$
7:             $S_{cds} \leftarrow S_{cds} \backslash CDS_k$
8:             $CDS_{ct} \leftarrow f_{\max}(S_{cds}, CDS_{ct})$
9:         **else if** $LT_v^u < \Theta$ & $CDS_{ct} isn't a backbone, v, u \subseteq V_{CDS_{ct}}$ **then**
10:            $\{CDS_k\} = \arg\{(v, u) \in E_{CDS_k}\}$
11:            $S_{cds} \leftarrow S_{cds} \backslash CDS_k$
12:            $CDS_{ct} \leftarrow f_{\max}(S_{cds}, CDS_{ct})$
13:         **end if**
14:     **else if** $N_2(v_i) \cap CDS_{ct} = \phi$ **then**
15:         $f_{\max}^w(N_1 v_m, N_1 v_m) \cap CDS_{ct} \neq \phi$
16:         $\{v_m\} \leftarrow \{v_m | N_1(v_m) \cap CDS_{ct} \neq \phi\}$
17:         $Domlist(v_i) \leftarrow f_{max}(W(N_1 v_m))$
18:     Update CDS_{ct}
19:     **end if**
20: **end if**

$ESm^{t_c}(S_{cds}) = \arg\max(Em^{t_c}(CDS_i))$. In order to reduce the renewal time of the backbone network, we use the parameter LT to describe the connection time of E_{cds_k}. The parameter of LT is estimated from current status of two nodes, which includes velocity, location and direction of movements.

In the previous section, we discussed the prerequisites of the maintenance algorithm. The nodes in FANET periodically exchange message to update the node information and determine whether the current network needs to update. In particular, we propose two kinds of maintenance mechanisms in this paper to maintain the network due to the different types of node failure.

Once the network changes, we need to determine whether the change in the FANET recurs a schedule. Typically, we consider the following three cases which is not necessary to carry out CDS scheduling algorithm.

- New nodes join the network.
  (a) $N_1(v) \cap V_{cds} = \phi$. Once the dominators around v detects a new node, the dominators will send a dominator packet message to the new node to identify its identity. Upon receiving the dominator messages, the new node updates $v.Domlist$ and sends a confirmation message to the node in the $v.Domlist$. The new node v may receive grouping messages from multiple dominators and select the neighbor node with the largest LT as the dominant point.
  (b) $N_1(v) \cap V_{cds} \neq \phi$. There are no dominators in the one-hop neighbor nodes of the newly joined node v. However, according to the characteristics of a connected graph, the two-hop neighbor nodes of v must have dominator nodes. If $N_2(v) \cap V_{cds} \neq \phi$, and $\{u \mid v \in N_1(u), N_1(u) \cap V_{cds} \neq \phi\}$, let dominatee u be the dominator of node v.
- Node v quits from the network and $v \notin V_{cds}$.
  Node quitting from the network will cause the network disconnected. It means there are no nodes in it's communication range. Nevertheless, the node could

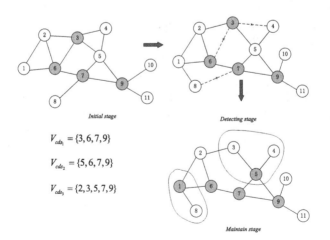

Fig. 7. An example of VBN maintenance.

readjust its motion state through the proposed SYN-*boid* motion rules based on the sensing data, since the sensing range R_e is larger than communication range R_c. After a certain time slots, it will detect the neighboring nodes. Then, according to the previous rule, it will be considered as a new node to join the network.

Typically, the current CDS requires to be updated under the following conditions.

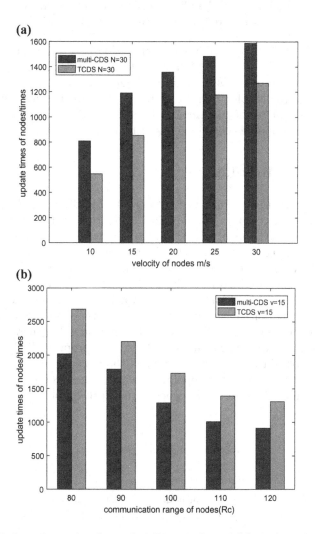

Fig. 8. (a). Update times of nodes with different velocity. (b). under different communication range.

- Changes in $E_{CDS_{ct}}$ or $V_{CDS_{ct}}$ occurred, where $E_{CDS_{ct}}$ is the edge set of current CDS and $V_{CDS_{ct}}$ is the node set respectively. In this case, there are two corresponding operations. Firstly, if node $v_i \in V_{CDS_{ct}}$ failed, the algorithm will delete $\{CDS_k\} = \arg\{v_i \in CDS_k\}$. Otherwise, if node $v, u \in V_{CDS_{ct}}$, $\{CDS_k\} = \arg\{(v, u) \in E_{CDS_k}\}$ isn't a backbone network, the algorithm will delete $\{CDS_k\} = \arg\{(v, u) \in E_{CDS_k}\}$, in which Θ is the link-up time threshold.
- The number of CDS in S_{cds} less than 2. In this case, the network topology has obvious changes, then the CDS construction algorithm will be used for re-searching S_{cds} until there are no CDSs in the graph.

An example of the VBN maintenance algorithm is represented in Fig. 7. Initially, we get the CDSs through MCDS construction algorithm. The CDS with largest weight is considered as the initial virtual backbone network. The second sub-graph shows the topology after a certain time slots, in which several edges are disconnected due to the movements of nodes. Finally, we maintain the VBN according to the proposed scheduling algorithm.

6 Simulation Results

In order to evaluate the performance of the Multi-MCDS Scheduling algorithm, we make comparative groups of numerical simulations with a typical maintenance algorithm based on single CDS. In the simulations, the topology samples are generated through the proposed SYN-*boid* model in this paper, with the parameters shown in Table 2. In the simulations, we assume that each node has the same communication radius. If the distance of two nodes are less than this threshold, they can communicate with each other. In order to make full analysis on the algorithm performances, we choose several scenes with different number of swarming UAV nodes. Initially, algorithm will randomly generate a group of nodes in a range of two-dimension space. Under the proposed SYN-*boid* model, algorithm will obtain a series of continuous topology snapshots with a fixed sampling rate. After getting the topology, we apply the Deep First Search algorithm in the simulations to detect whether network topology is connected at each time slot.

Table 2. Simulations parameters.

Parameter	Parameter description	Value
N	Number of nodes	20,30, ... 80
T	Total time slots	500
τ	Time slot interval	3 s
v	Velocity of nodes	10/20/30 ... m/s
R_c	Communication radius	80/90/100 ... 120 m

The maintenance algorithm based on single CDS is a general method in topology maintenance of MANETs. It maintains the network by selecting the backup nodes. However, the backup nodes in this algorithm are not always existed. Therefore, this may cause unexpected failures of network maintenance. In the maintenance phase, we hope to achieve a less update overhead and larger success probability. In the simulations, we obtain a series of continuous topological snapshots under the proposed SYN-*boid* model. We simulated 500 time slots in different scenarios and analyzed the impacts of different parameters on performance.

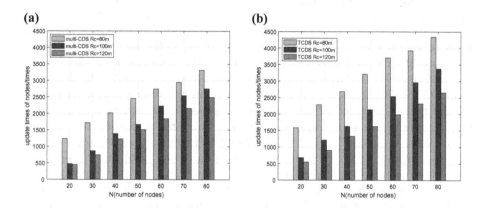

Fig. 9. (a). Relationship between T_{cost} and N, R_c of Multi-CDS. (b). Relationship between T_{cost} and N, R_c of TCDS.

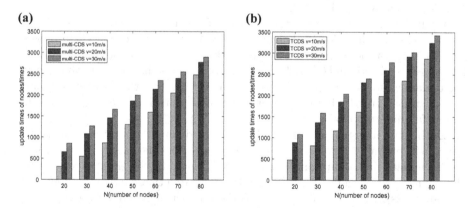

Fig. 10. (a). Relationship between T_{cost} and N, v of Multi-CDS. (b). Relationship between T_{cost} and N, v of TCDS.

As shown in Fig. 8(a), for a more comprehensive comparison, we set different scenarios with a velocity of 10 to 30. Here we choose the number of nodes as 30 and the communication radius as 100. Generally, the typical CDS updating algorithm (TCDS) updates the backbone network through detection of the network periodically, if any node fails or affect the current, the network will reconstruct the CDS and update nodes's. Under different conditions of velocity, the typical CDS updating algorithm (TCDS) takes higher update overheads than the proposed Multi-MCDS algorithm all the way. The results show that the increasing of velocity will cause an increase in update overheads. The reason for this situation is that the connectivity between nodes changes more frequently. In order to maintain the VBN in the network, more nodes need to be scheduled. Moreover, it can be easily observed the T_{cost} in TCDS increases greater than in the proposed algorithm. We can know that lower velocity will get smaller update overhead.

The result in Fig. 8(b) shows the relationship between update overheads and communication radius of the two algorithms. In the simulation, we set different communication radius, ranging from 80 m to 120 m. And the number of nodes in the scenarios is 40. In the contrast, the proposed algorithm performs better than TCDS. We can see that the increase of communication radius will result in smaller update overhead in the maintenance phase. Moreover, with the increase of communication radius, the update overhead becomes more stable.

We have compared the effects of velocity and communication radius on the update overhead earlier. In the simulations of Fig. 9(a) and (b), the scenarios are generated under both different number of nodes N and communication radius. Figure 9(a)shows the relationship between nodes number and update overhead in the proposed algorithm. Figure 9(b) shows the relationship between nodes number and update overheads in TCDS. It can be seen that the proposed algorithm has less update overheads under the same condition of nodes number and velocity. As the nodes increases, the update overheads increase. This is because the increase of nodes will cause more topological changes. We need to schedul more CDS nodes to maintain the normal operation.

Figure 10(a) and (b) shows the changes of update overhead when the number of nodes and the velocity change. It is obviously indicated that the proposed algorithm has a more stable performance than TCDS, and the updating overheads are smaller than TCDS. The velocity of the node has a great influence on the update costs. Moreover, with the increase of velocity, the successful update rate decreases greatly. It means that the velocity has a greater impact on the successful update probability.

During the maintenance phase, due to high dynamic of nodes, the maintenance will probably fail. Figure 11(a) represents the relationship between successful update probability and the velocity of node. In the simulations, we set the communication radius Rc by 100 m. It is indicated that the successful update probability is not necessarily related to the number of nodes. It is noted that, both the proposed algorithm and TCDS perform well when the velocity of nodes is small. However, the TCDS has poor performance when the velocity increases. It is because the high speeds of nodes cause TCDS algorithm cannot find stable backup nodes to maintain VBN. The performance of the proposed Multi-CDS algorithm is better than TCDS in the condition of high speed.

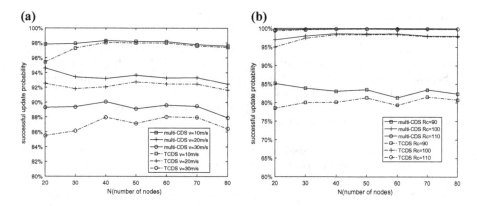

Fig. 11. (a). Successful updates probability under different velocity. (b). Successful updates probability under different communication radius.

The results in Fig. 11(b) show the influence of the changes of communication radius on the success probability of the node updating. With the increase of communication radius, the successful update probability rises up. Moreover, with the same velocity, the increase of node number will lead to the increase of successful update probability. The reason is that, the scheduling of CDSs can repair the VBN effectively since the connectivity among nodes is strengthened. The simulation indicates that both algorithms perform well when the communication radius is large.

7   Conclusion

In this paper, we proposed a multi-MCDS algorithm to construct and maintain virtual backbone network by scheduling multiple CDSs in the network. In particular, we compared the impacts of node velocity and communication radius on the performance of the algorithm. The simulations results show that, the update overheads of the proposed algorithm is much less than TCDS, and the successful update probability is higher than TCDS with the increases of nodes. Moreover, the proposed algorithm can solve the failures of topological changes caused by high dynamics of nodes.

Acknowledgment. The authors would like to express their high appreciations to the supports from the National Natural Science Foundation of China (61871426), National Science and Technology Major Project (91538110) and Basic Research Project of Shenzhen (JCYJ20170413110004682).

References

1. Zhong, D., Zhu, Y., You, T., Kong, J.: Topology control mechanism based on link available probability in aeronautical ad hoc network. J. Netw. **9**(12) (2014)
2. Sahingoz, O.K.: A delay-aware stable routing protocol for aeronautical ad hoc networks. J. Intell. Rob. Syst. **74**(1–2), 513–527 (2017)
3. Gu, W.-z., Li, J.-l., He, F.-j.: Topology control mechanism based on link available probability in aeronautical ad hoc network. J. Inf. Comput. Sci. **9**(2), 347–359 (2012)
4. Lin, Y.K., Huang, C.F.: Assessing reliability within error rate and time constraint for a stochastic node-imperfect computer network. Proc. Inst. Mech. Eng. Part O J. Risk Reliab. **227**, 80–85 (2013)
5. Yan, Z., Nie, C., Dong, R., Gao, X., Liu, J.: A novel OBDD-based reliability evaluation algorithm for wireless sensor networks on the multicast model. Math. Probl. Eng. **2015**(3), 1–14 (2015)
6. Kuo, S.Y., Yeh, F.M., Lin, H.Y.: Efficient and exact reliability evaluation for networks with imperfect vertices. IEEE Trans. Reliab. **56**(2), 288–300 (2007)
7. Du, H., Wu, W., Ye, Q., Li, D., Lee, W., Xu, X.: CDS-based virtual backbone construction with guaranteed routing cost in wireless sensor networks. IEEE Trans. Parallel Distrib. Syst. **24**(4), 652–661 (2013)
8. Deng, C., Gao, F., Zhao, L.F.: A topology maintenance algorithm used for wireless sensor networks. In: Fourth International Conference on Multimedia Information Networking and Security, pp. 168–171 (2013)
9. Manikonda, A.: A new method for controlling and maintaining topology in wireless sensor networks. Int. J. Comput. Netw. Commun. **6**(4), 91–98 (2014)
10. Zou, F., et al.: New approximations for minimum-weighted dominating sets and minimum-weighted connected dominating sets on unit disk graphs. Theoret. Comput. Sci. **412**(3), 198–208 (2011)
11. Dai, F., Wu, J.: On constructing k-connected k-dominating set in wireless networks. In: IEEE International Parallel and Distributed Processing Symposium, 2005, Proceedings, p. 81.1 (2004)
12. Wu, Y., Wang, F., Thai, M.T., Li, Y.: Constructing k-connected m-dominating sets in wireless sensor networks. In: Military Communications Conference, 2007, Milcom, pp. 1–7 (2007)
13. Liao, Y., Qi, H., Li, W.: Load-balanced clustering algorithm with distributed self-organization for wireless sensor networks. IEEE Sens. J. **13**(5), 1498–1506 (2013)
14. Wang, J., Kodama, E., Takata, T.: Construction and maintenance of K-hop CDS in mobile ad hoc networks. In: IEEE International Conference on Advanced Information Networking and Applications (2017)
15. Reynolds, C.W.: Flocks, herds and schools: a distributed behavioral model. In: Conference on Computer Graphics and Interactive Techniques, pp. 25–34 (1987)
16. Fu, D., Han, L., Yang, Z., Jhang, S.T.: A greedy algorithm on constructing the minimum connected dominating set in wireless network. Int. J. Distrib. Sens. Netw. **12**(7), 1703201 (2016)

A Novel Topology Design Method
for Multi-layered Optical Satellite Networks

Xiupu Lang[1], Qi Zhang[1], Lin Gui[1(✉)], Xuekun Hao[2], and Haopeng Chen[3]

[1] Department of Electronic Engineering,
Shanghai Jiao Tong University, Shanghai 200240, China
guilin@sjtu.edu.cn
[2] The 54th Research Institute of China Electronics Technology Group Corporation,
Shijiazhuang 050000, Hebei, China
[3] Department of Software Engineering, Shanghai Jiao Tong University, Shanghai 200240, China

Abstract. Recently the topology design of multi-layer satellite network has drawn much attention from researchers due to the application of laser link in inter-satellite links (ISLs). Snapshot division is a common method of dynamic topology design, which splits the time-varying topology into a sequence of static snapshots. However, previous topology design researches based on snapshot division seldom consider multi-objective optimization and practical link parameters, such as link capacity, degree constrains, and link switch time. In this paper, a novel topology design method aiming to concurrently minimize both link switch time and network average end-to-end delay is proposed. The constraints of degree, visibility, and connectivity are considered in the proposed topology design method. Simulation results show that the proposed strategy has a better performance compared to traditional topology design methods for multi-layered optical satellite networks.

Keywords: Topology design · Optical satellite networks · Non-equal length snapshot division

1 Introduction

The noticeable trend in multi-layer satellite network is the application of laser link in satellite networks [1]. The network based on laser link has prominent superiority to traditional networks based on microwave communications, especially in broadband use, such as reconnaissance, emergency communication, earth observation, etc. [2, 3]. Though link capacity is drastically boosted, laser link demands higher link reliability and longer link setup time. Moreover, a self-organized heterogeneous satellite network without terrestrial control is expected. In such conditions, the significance of topology design for a reliable and efficient network is highlighted.

This work is supported in part by the National Natural Science Foundation of China (61671295, 61420106008, 61671128), and the Shanghai Key Laboratory of Digital Processing; it is also partly sponsored by (JMRH-2018-1074).

Optical heterogeneous satellite network features for these following characteristics:

(1) Mobility: satellite orbital operation would lead to changes in network topology and link continuity. Nevertheless, mobility of satellite is predictable and periodic in contrast to random movement. When it comes to a heterogeneous network, different layers of satellites have different periods and network period is the least common multiple of different satellite periods.

(2) The number of antennas on the satellite is limited due to the weight and energy constraints of onboard payload. Therefore, it often occurs to one satellite that the number of visible ones is greater than the number of connectable ones.

(3) Satellite network is a resource-constrained system since its energy consumption, band allocation is strictly limited.

In such scenarios, there exist several problems to be solved in the field of topology design: (1) how to measure and guarantee the reliability of the network. (2) how to determine the topological connection between satellite such that throughput is maximized or network average delay is minimized. (3) based on snapshot division, how to determine the number of snapshots so as to save storage resources as much as possible. Large variety of works are conducted to find solutions to the above problems. Zheng et al. in [4] took the eigenvalue of the Laplacian matrix corresponding to the topology as the connectivity metric and guarantees the reliability of the network by maximizing connectivity. In [5], the authors aimed to reduce the number of snapshots under the constraint of time delay by optimizing equal length snapshot division. The system period duration may be long (even up to 7 days) and storing all the snapshots in one period consumes a large bulk of storage resources. Li et al. minimized the network cost under the constraint of reliability [6]. Its main innovation lies in the modeling of practical link characteristic in contrast to ideal links in many other works.

As far as we know, few studies have been conducted for optimizing network delay and rarely considering the laser link. In optical satellite networks, link switch time cannot be ignored. In this paper, we propose a topology design strategy based on non-equal length snapshot division with the joint consideration of time delay and network reliability. Most studies only consider equal-length slot division, which is impractical [7]. The time delay consists of two parts: link switching delay and transmission delay. The objective in the current work is to minimize the overall delay. Besides, we ensure the reliability of the network by ensuring that the network is fully connected and the degree of each satellite node is maximized. Simulation results reveal that our strategy achieves a better performance compared with the longest visibility strategy.

2 System Model

The problem undertaken in this paper is to find near-optimal topology design which achieves as small a delay as possible and satisfies reliability and connectivity constraints. The following aims to mathematically formulate a model of the topology design problem.

Snapshot: a snapshot refers to a period when the topology connection remains fixed. If the network topological connection is represented as a matrix. In one snapshot, the matrix is unchanged.

Network period: For the heterogeneous network consisting of LEO, MEO, GEO, the network period is the least common multiple of all satellite periods.

2.1 Visibility, Distance and Connection Matrix

Since the mobility of satellites is periodic and predictable, the trajectory of the satellite and the relative distance between them are significant inputs to the model. These data could be easily exported from the software named STK, therefore, we just take them as algorithmic parameters.

After deriving the distance between every satellites, next we would build the distance matrix and visibility matrix. Suppose that every satellite is denoted as a node and every ISL is denoted as an edge. Then satellite network can be represented as an undirected graph $G = (V, E)$, where $V = \{1, 2, \ldots, M\}$ is the vertices set of satellite nodes and $E = \{e_{ij} | i, j \in V, i \neq j\}$ is the edges set of ISLs. Let binary variable a_{ij} denotes the visibility between any two satellites, i.e. if v_i is visible to v_j, $a_{ij} = 1$, if v_i is not visible to v_j, $a_{ij} = 0$. Since the satellite network has no self-loop, $a_{ii} = 0, i = 1, 2, \ldots, M$. Hence, a symmetric $M \times M$ matrix $A \triangleq \{a_{ij}\}$ is defined as the visibility matrix. The symmetric $M \times M$ distance matrix $D \triangleq \{d_{ij}\}$ is defined as follows:

$$d_{ij} = \begin{cases} 0, & a_{ij} = 0 \\ \text{distance between } v_i \text{ and } v_j, & a_{ij} \neq 0 \end{cases} \tag{1}$$

Besides, a symmetric $M \times M$ matrix $H \triangleq \{h_{ij}\}$ is used to represent the connection relationship between v_i and v_j. If there exists a link between v_i and v_j, $h_{ij} = 1$, otherwise, $h_{ij} = 0$.

2.2 Average Time Delay

The current work uses average time delay as the objective function. The time delay is comprised of two parts: average transmission delay and average link switch delay. According to [8], link switch time could be formulated as follows:

$$t_{link\_switch} = 2(t_s + t_L) + t_r \tag{2}$$

where t_s is the duration for adjusting angle of view, t_L is path transmission time, t_r is the conversion time between the laser and the detector. When the farthest communication distance is 70000 km and link switch is needed, $t_{link\_switch} = 2$ s, otherwise, $t_{link\_switch} = 0$ s. Then we could derive average link switching delay by the following formula:

$$T_{link} = \frac{1}{M} \sum_{i=1}^{m} t_{ij} \tag{3}$$

where M is the total number of links and t_{ij} is the link switch time between v_i and v_j.

As for transmission delay, we can adopt the M/M/1 queuing model to obtain this value [9]. Assume that the arrival of each service obeys the Poisson distribution, there are a total of W businesses in the network and the arrival of W kinds of business is independent of each other, then one of the generally accepted formulation which is accurate enough for the measurement of network average delay is:

$$T_{transs} = \sum_{l=1}^{m} \frac{f_l}{C_l - f_l} / \sum_{i=1}^{W} \lambda_i \tag{4}$$

where m is the number of links in the network, C_l denotes lth link's capacity, f_l denotes the load of lth link. We can use Shannon's formula to estimate the upper bound of the laser channel capacity. Meanwhile, the path loss and transmission distance are logarithmically related. Thus, we can substitute transmit power, antenna gain, and distance into Shannon formula to obtain the capacity for every link.

According to formula (3) and (4), we can derive average time delay as follows:

$$T_i = \frac{1}{M} \sum_{i=1}^{m} t_{ij} + \sum_{l=1}^{m} \frac{f_l}{C_l - f_l} / \sum_{i=1}^{W} \lambda_i \tag{5}$$

T_i denotes the average time delay for snapshot i.

2.3 Constraints

Considering the limitation of the number of laser terminals on the satellite, it is assumed that a satellite has a degree constraint d_i. The value d_i equals to the number of laser terminals on satellite v_i. Hence, connection matrix must satisfy the degree constraint, i.e. $\sum_{j=1}^{M} h_{ij} \leq d_i$, $i = 1, 2, \ldots, M$. For simplicity, the constraint could also be written in the matrix form:

$$A1 \leq d \tag{6}$$

where $1 = \{1, 1, \ldots, 1\}^T$ is a $M \times 1$ vector.

Connection matrix must satisfy the visibility constraint. In matrix form, the constraint can be expressed as:

$$H \leq A \tag{7}$$

In order to use formula (4) to calculate average transmission delay, the link must satisfy flow constraint. It can be expressed as follows:

$$f_i \leq C_i, i = 1, 2, \ldots, M \tag{8}$$

In addition, the network must be fully connected. Full connectivity can be easily checked by using BFS for each node. Let matrix G represents the outcome of BFS. If v_i could find v_j using BFS, $g_{ij} = 1$, otherwise $g_{ij} = 0$. Besides, we define $g_{ii} = 1, i = 1, 2, .., M$. The constraint must satisfy:

$$G = 11^T \tag{9}$$

2.4 Problem Formulation

The topological design problem aims to achieve two minimization objectives related to average transmission delay and link switch delay. The optimization problem can be formulated as follows:

$$
\begin{aligned}
\min \ & \frac{1}{T}\left(\sum_{i=1}^{k} (T_{link}^{(k)} + T_{trans}^{(k)}) \right) \\
\text{s.t.} \ & A1 \leq d \\
& H \leq A \\
& f_i \leq C_i, i = 1, 2, \ldots, M \\
& G = 11^T
\end{aligned} \tag{10}
$$

where k is the number of snapshots, T is the network period.

The difficulty in solving optimization problem (10) lies in how to properly divide duration T into a sequence of snapshots. Since the start and end time of one snapshot can be randomly and continuously selected in the duration T, the number of snapshots is unknown and the duration of each snapshot is also unknown. In theory, the combination of the number of snapshots and the length of snapshots is infinite. The solution space is infinite. Hence, problem (9) is a NP-hard problem.

3   Minimum Delay Topology Design

Due to the NP-hard nature of the problem (10), we propose a heuristics called minimum delay topology design (MDTD) to construct a topology. One of the conventional principles of topology design is linking to the longest visible satellite. Though it could achieve the highest reliability, this principle is not efficient enough because link capacity is inversely proportional to distance. Although two satellites are visible, the distance between them is too long to be suitable as a transmission link. The algorithm MDTD aims to find the most suitable link switching time to reduce overall network latency. However, the price is the decrease in network stability caused by the increase in the number of link switching times.

One of the intuitive ideas to construct a low-delay and fully connected graph is to firstly find the shortest path for every node pair v_i and v_j using Dijkstra's algorithm, the shortest path is denoted as $P^G_{\text{dis}}(v_i, v_j)$. Then the union of all of them must be the optimal solution with the smallest delay for one snapshot. However, this solution might not satisfy the constraints. Then based on the greedy algorithm, add and delete edges to make the previous optimal solution satisfy the constraints. The eventual solution is suboptimal.

The above is the complete process of obtaining a topology design for a single snapshot. After setting a network status sampling interval T_s to update distance for every link, we apply the above topology design method at each sampling time point and we can obtain a topology for every sampling point. Assume that the connection matrix at sampling point i-1 is denoted as $H^{(i-1)}$ and the distance matrix at sampling point i is denoted as $D^{(i)}$. After applying the above topology design method we obtain a new connection matrix $H^{(i)}$. If $H^{(i-1)} \neq H^{(i)}$, we should devise a metric to decide whether to change link connection since the link switch increases network latency. The method is as follows. Assume the average transmission delay for sampling time point i-1 under the topology $H^{(i-1)}$ and the distance matrix $D^{(i)}$ is T_{old}, the average transmission delay for sampling time point i under the topology $H^{(i)}$ and the distance matrix $D^{(i)}$ is T_{new}, if and only if delay satisfies the following formula (11), switch occurs.

$$T_{new} + T_{link} < kT_{old} \tag{11}$$

where k is a factor which reflects the designer's preference.

The algorithm is shown as follows:

Algorithm Minimum Delay Topology Design

Input:D, T_s, T, H

1: $T_{now} = T_s \quad S = \left\{ \left(v_i, v_j \right) \right\}, 1 \leq i, j \leq M, i \neq j$

2: former_H = H

3: **while** $T_{now} < T$ **do**

4: Updata D, T_{now}

5: Calculate transmission delay T_{old} using former_H

6: **for all** pairs $\left(v_i, v_j \right) \in S$ **do** ,

7: $H \leftarrow \phi$,

8: Find the shortest path $P_{dis}^G(v_i, v_j)$ using Dijkstra's method

9: **if** $e \in P_{dis}^G \left(v_i, v_j \right)$ and $e \notin H$ **then**

10: $H = H \cup \{e\}$

11: **end if**

12: **end for**

13: **for all** node v_i **do**

14: **if** degree(v_i) > degree_constrain **then**

15: Sort all links connected to v_i in set   according to their distance

16: **for all** $e \in \varepsilon$ (processed in decreasing order of costs) **do**

17:          **If** delete e, network is fully connected **then**

18:          $$H=H\text{-}\{e\}$$

19:          **end if**

20:          **end if**

21:     **end for**

22:     **for all node** v_i **do**

23:          **if** degree(v_i) < degree_constrain **then**

24:               Sort all links connected to v_i in set ε according to their distance

25:               **for all** $e \in \varepsilon$ (processed in increase order of costs**) do**

26:                    **if** add e, all node still satisfy degree_constrain, **then**

27:                         $$H=H\text{+}\{e\}$$

28:                    **end if**

29:               **end for**

30:          **end if**

31:     **end for**

32:     Calculate transmission delay T_{new} using H and **link switch time** T_{link}

33:     **if** $T_{new} + T_{link} < kT_{old}$ **then**

34:          former_H = H

35:     **end if**

36: **end while**

In MDTD, the algorithm firstly finds the shortest path for each pair of node pairs. Its time complexity is $O(n^2(m + n)logn)$ since $O(n^2)$ times of Dijkstra's algorithm whose time complexity is $O((m + n)logn)$ needs to be done. Secondly, the algorithm adds or deletes edges to satisfy the constraints. In this step the time complexity is $O(n^2)$ since the algorithm would traverse each edge of each node. Hence, the total time complexity of MDTD is $O(n^2(m + n)logn + n^2)$.

4 Simulation

4.1 Simulation Environment

We select a satellite network composed of 6 MEOs and 3 IGSOs. 6 MEOs are evenly distributed on three orbital planes that are 120° apart. The detailed orbital parameters of the satellite are shown in Table 1 below and satellite constellation is shown in Fig. 1. The network period is one solar day and the simulation time one network period. In this constellation, IGSO satellites are permanently visible to each other during a network period. MEO1 and MEO2, MEO3 and MEO4, MEO5 and MEO6 are respectively located on the same orbital plane. Satellites located in the same orbital plane are permanently visible. Therefore, link switch can only occur between MEO and IGSO or between MEO belonging to different orbital plane.

Table 1. Orbital parameters of each Satellite

	Semimajor axis	Eccentricity	Inclination	Argument of perigee	RAAN	True anomaly
IGSO1	42157 km	0.0031	56.07°	186.40°	68.12°	55.165°
IGSO2	42157 km	0.0072	54.15°	232.87°	189.69°	55.1
IGSO3	42157 km	0.0066	52.59	201.06°	305.68°	55.17°
MEO1	27906 km	0.0007	55.07°	4.93°	39.64°	332.79°
MEO2	27906 km	0.0068	55.09°	4.93°	40.12°	62.79°
MEO3	27906 km	0.0048	55.00°	290.41°	160.11°	284.93°
MEO4	27906 km	0.0011	55.01°	272.68°	158.42°	14.93°
MEO5	27906 km	0.0007	55.09°	121.65°	281.63°	46.43°
MEO6	27906 km	0.0073	55.08°	121.46°	279.86°	135.83°

Other simulation parameters are set as follows: The laser link bandwidth is 1G, transmit power is 105 dBW, antenna gain is 109 dB. Assume that there are two kinds of services in the network: multimedia service and speech service. The two services are considered to be independent and the traffic of every node in the network is subject to Poisson distribution [10]. The traffic model is shown in Table 2.

Table 2. Traffic model of computer and mobile phone

	Single-Channel data	Average arrival rate λ
Multimedia service	2 Mbps	50
Speech service	64 kbps	1536

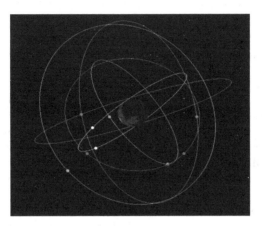

Fig. 1. Satellite constellation

4.2 Results and Analysis

The Minimum Delay Topology Design (MDTD) strategy divides and merges snapshots with the sampling interval as the minimum granularity. And based on snapshot division, MDTD the uses greedy algorithm to design the connection relationship. In this section, we would give the topological design result of the algorithm. Furthermore, to check the efficiency of the proposed MDTD, its performance is compared with the Maximum Visibility Topology Design (MVTD), in which one satellite is always connected to other satellites with the longest visible time.

We set $k = 1$ in MDTD. The network sampling time ranges from 5 min to 60 min and the step is 5 min.

Figure 2 shows the relationship between the average network delay and the sampling time concerning MDTD and MVTD. We could derive from Fig. 2 that the network average delay could deteriorate if the sampling interval is too small or too large with regard to MDTD. The best sampling interval for MDTD is approximately 20 min. We can also derive that delay derived by MVTD is nearly the upper bound of the delay derived by MDTD.

Relationship between the link switch time and the sampling time for MDTD and MVTD is shown in Fig. 3. According to Fig. 3, in MDTD as the sampling interval increases, the number of link switches decreases and eventually converges to the number of link switches of the MVTD algorithm.

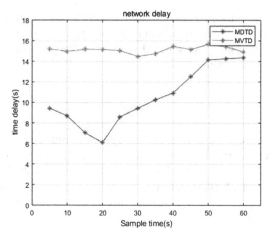

Fig. 2. Network average delay of MDTD and MVTD

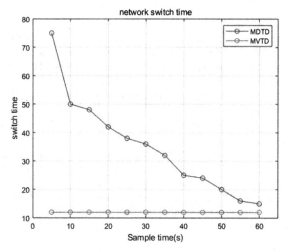

Fig. 3. The number of link switches of MDTD and MVTD

The results of the two algorithms with a sampling interval of 20 min are listed in Table 3 below:

Table 3. Performance comparison with $T_s = 20$ min

	Network average delay	The number of link switches
MDTD	6.103 s	42
MVTD	15.154 s	12

According to Table 3, the MDTD algorithm has a 59% improvement over the MVTD algorithm in network average delay, which is attained by increasing the number of link switches from 12 to 42. This result shows that the MDTD algorithm is efficient and can be applied in a practical system.

5 Conclusion

In this paper, we have proposed an effective strategy to determine the connection relationship with the joint consideration of snapshot division and network delay. The proposed MDTD algorithm made a trade-off between link switch and link stability. According to the simulation results, the MDTD algorithm has been confirmed to provide a noticeable improvement with respect to network average time delay. Meanwhile, network stability is also greatly reduced.

References

1. Fernández, F.A.: Inter-satellite ranging and inter-satellite communication links for enhancing GNSS satellite broadcast navigation data. Adv. Space Res. **47**(5), 786–801 (2011)
2. Xu, Y., Chang, Q., Yu, Z.J.: On new measurement and communication techniques of GNSS inter-satellite links. Sci. China Ser. E: Technol. Sci. **55**(1), 285–294 (2012)
3. Chan, V.W.S.: Optical satellite networks. J. Lightw. Technol. **21**(11), 2811–2827 (2003)
4. Zheng, Y., et al.: Weighted algebraic connectivity maximization for optical satellite networks. IEEE Access **5**, 6885–6893 (2017)
5. Huang, J., et al.: An optimized snapshot division strategy for satellite network in GNSS. IEEE Commun. Lett. **20**, 2406–2409 (2016)
6. Li, F., Chen, S., Huang, M., et al.: Reliable topology design in time-evolving delay-tolerant networks with unreliable links. IEEE Trans. Mob. Comput. **14**(6), 1301–1314 (2015)
7. Chang, H.S., Kim, B.W., Lee, C.G., et al.: FSA-based link assignment and routing in low-earth orbit satellite networks. IEEE Trans. Veh. Technol. **47**(3), 1037–1048 (1998)
8. Zhongke, J.: Studies on some key technologies of inter-satellite laser communication (2017)
9. Yongli, X.: Calculation of average delay in computer networks. J. Shijiazhuang Univ. Econ. **2**, 208–212 (1998)
10. Guoli, W.: Research on spatial optical network routing and resource management technology (2018)

An SDN-Based Dynamic Security Architecture for Space Information Networks

Ziqi Wang[1,2(✉)], Baojiang Cui[1,2], Shen Yao[1,2], and Meiyi Jiang[1,2]

[1] School of Cyberspace Security, Beijing University of Posts
and Telecommunications, Beijing, China
wangziqi@bupt.edu.cn
[2] National Engineering Laboratory for Mobile Network Security, Beijing, China

Abstract. In the near future, the Space Information Network (SIN) will evolve into Space-earth Integrated Network, which will realize the interconnection among the Space-based Network, traditional Internet and Mobile Communication Network. The construction of the future SIN will provide the ability of global real-time communication and comprehensive information services, which makes it possess a wide application prospect. The future SIN will realize inter-satellite routing and its network capacity will be flexible, its network topology and link status will change dynamically as well, which will increase the complexity of the network management. In addition, the openness of communication link and the resources limitation of the devices will both bring security threats to the SIN. We study on the characteristics of future SIN and present a dynamic security protection architecture based on SDN (Software Defined Network). We study on the trusted authentication mechanism of dynamic networking entities and the technique of path optimization and risky-path isolated transmission. At last, we design a network intelligent security management mechanism at the top level. Our scheme can greatly improve the security of future SIN.

Keywords: Space information network · Software defined network · Dynamic security architecture

1 Introduction

Nowadays, traditional Bent Pipe satellite communication system is faced with puzzles. For example, the communication satellites need the cooperation of ground stations, therefore, the cost of constructing adequate ground stations will be extremely high. At the same time, with the development of satellite communication technology and the ever increasing demand for satellite network services, many countries have proposed their SIN construction plan, such as the *Starlink* plan of *SpaceX*, aiming to provide global real-time broadband Internet services.

According to the services requirements, the future SIN will realize the inter-satellite link and the satellite will deploy corresponding payloads. In addition, SIN will have the inter-satellite networking capabilities and allow user terminals to dynamically access. More importantly, the future SIN will evolve into the Space-earth Integrated Network,

© Springer Nature Singapore Pte Ltd. 2020
Q. Yu (Ed.): SINC 2019, CCIS 1169, pp. 99–111, 2020.
https://doi.org/10.1007/978-981-15-3442-3_9

which will be integrated with the traditional Internet and mobile communication network, to provide users with communication and internet access services [1].

Compared with the traditional terrestrial network, the topology of the future SIN is highly complex and dynamic. Even worse, due to the openness of its communication links and the inherent long communication delay, the traditional network security management methods are not applicable for use. Considering the particularity of SIN, centralized management and control capabilities are needed. Data leakage, data tampering, denial of service attacks and other unsafe factors may greatly threat the network security, therefore, the adaptive adjustment of security capabilities of SIN is also necessary [2]. For the large number of users accessing to SIN, access control mechanism is extremely essential to prevent the unauthorized accessing, and real-time optimization of routing is highly required to provide high quality communication service.

In this paper, we analyze the characters and security requirements of future SIN, and propose a scheme of applying SDN to SIN. Since the control plane and data place are suggested to be decoupled, SDN architecture provides flexible network management capabilities and enables administrator to manage network services and security by programing [3, 4]. Based on this architecture, we designed a trusted authentication mechanism, a dynamic routing optimization mechanism and an intelligent security management mechanism. In summary, the main contributions of our paper is listed as follows:

- An SDN-based SIN security protection architecture. Based on this architecture, we realize a dynamic adjustment method of security component and increase the security ability of the future SIN.
- A trusted authentication mechanism for network entities. In this way, we can identify the legitimacy of the terminal and achieve a mutual authentication between nodes.
- A routing optimization and risk isolation mechanism. It provides a real-time method of routing optimization and isolates the security risks in the complex and risky network environment of the future SIN.
- An intelligent security management mechanism. With this method, we can detect the security threats timely and adopt the corresponding security strategies to deal with them.

The remaining parts of our paper are organized as follows. In Sect. 2, we describe the network architecture and our proposed SDN-based security protection framework. In Sect. 3, we describe the specific techniques applied in our model and evaluated them by demonstration and analysis. The last section is a conclusion and prospect.

2   SDN-Based Security Protection Architecture for SIN

In the future, SIN will consist of satellites, terrestrial gateways and mobile terminals, and the satellite network can be divided into three layers according to the orbit: the low earth orbit layer (the LEO layer), the medium earth orbit (the MEO layer) and the geosynchronous earth orbit (the GEO layer), all of which can achieve global coverage, and the higher-layer orbit can cover the lower one sequentially. At the same time, the SIN constitutes an independent spatial autonomous system, which is automatically networked and operated in space; the ground gateway is directly connected to the visible satellite, responsible for address translation and communication between the satellite network and the terrestrial network; the terminal is connected to the visible satellite and the communication between terminals is forwarded by satellites.

2.1 SDN Applied to SIN

According to the characteristics of the above-mentioned future SIN and the orbit theory of the spacecraft, the constellation topology of SIN will change dynamically, and the quality and security of the communication link will change accordingly. The fundamental idea of SDN is to separate the control functions from the network devices and concentrate them in a centralized controller. Then, the data forwarding is conducted by the flow table distributed by the SDN controller. In this way, the administrator realizes flexible management of network and dynamic services optimization in a low cost and adaptable way [5]. We virtualize the network function of SIN infrastructure based on SDN to set up a resources pool, which consists of forwarding components and security components. We establish a control layer above the forwarding layer and utilize flow table to manage the traffic forwarding and resources scheduling [3, 4], aiming to provide a dynamic reorganization capability of security functions and rich APIs for the application layer on the top of the architecture. Considering that the future SIN will provide diversified services, we set up an application layer to analyze the demands of users and combine the service components and the security components with the help of APIs provided by the control layer. The overall SDN-based technical architecture is depicted in Fig. 1.

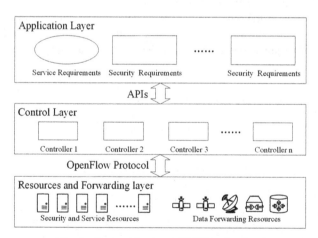

Fig. 1. The architecture of SDN technology

2.2 Threats Analysis and Security Requirements

In the future SDN-based SIN architecture, satellite nodes will be faced with dynamic networking requirements and massive user terminals will also dynamically access and leave the network, which will bring weak points to the networking and accessing levels. Therefore, trusted authentication mechanism for network entities is essential.

The openness of inter-satellite communication link will lead to diversification of attack types and security issues, which may cause severe consequences. Attackers can launch malicious attacks in any way, anytime and anywhere. Therefore, it is necessary to identify and classify the attack modes and use the corresponding security strategies to mitigate the threats.

As for the communication and network services quality, since the topology and security situation will be time-varying, it is essential to optimize the routing in real time and manage the security behaviors globally. Hence optimizing the route according to the quality parameters of the inter-satellite links and isolating the threats according to the security parameters can effectively improve the service quality and security capabilities of the system.

2.3 Network Architecture

The core part of the overall network architecture consists of a ground service center and a satellite-based network.

The SDN controller is deployed in the main ground service center and is responsible for the centralized control and management of the entire network. There are also distributed services centers deployed on other suitable ground regions. Each service center can communicate with the satellites through the ground gateway and base station to provide users the ability of accessing the broadband Internet.

In addition to functioning as a controller, the service center will also function as the authentication center to verify the identity of the terminal and issue a certificate to it, as the training center of the intelligent anomaly detection to obtain an optimal training models and deliver the optimized model to the satellite nodes, and as a routing calculate center and flow table deliver center to manage the traffic forwarding of the entire network.

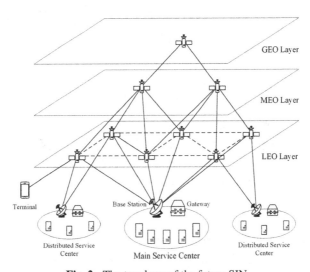

Fig. 2. The topology of the future SIN

Considering that the compute resources and storage resources of the satellite is extremely limited and the inherent time delay caused by inter-satellite communication link, most of the computing tasks are assigned to the services center. However, a data synchronization mechanism between satellite and ground service center is proposed to provide authentication function with a shorter time delay if needed.

The physical topology is shown in Fig. 2.

3 Proposed Techniques Based on Our Architecture

Combined with the SDN-based SIN architecture, the relationship between our proposed security techniques can be shown as below (Fig. 3).

At the bottom of architecture, we decouple the control functions from the devices to form the resources and data forwarding layer and provide dynamic adjustment capabilities for the upper layers. In the middle layer, which is the control layer, we apply our proposed trusted authentication mechanism and routing optimization mechanism to provide management and control basis for SDN controller. At the top is the application layer, we apply the proposed intelligent security management mechanism to identify the abnormal traffic and analyze the appropriate management strategy autonomously, and then call the lower control layer through the APIs to realize dynamic control of the security services.

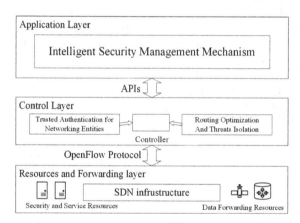

Fig. 3. Security protection mechanisms base on the SDN architecture and the relationship between them

3.1 Trusted Authentication Mechanism for Dynamic Networking Entities

Requirements. In the future, the topology of SIN will change dynamically, and the terrestrial network and the satellite network will be dynamically integrated. The access of the mobile terminal will greatly affect the security of the overall network [6]. The trusted authentication mechanism for dynamic networking entities will strictly authenticate the identities of the entities to ensure the security of the entities which intend to join the network. Besides, so as to resist tampering and replay attacks we introduce timestamp mechanism to check the integrity of the communication traffic.

Related Work. In the existing researches of SIN authentication techniques, researchers mainly study on authentication algorithms and authentication algorithms applicable to different network models. Zhu et al. [6] proposed an inter-satellite network authentication scheme suitable for double-layer satellite networks. Take the advantages of the predictability of the satellite orbits, they introduce a pre-authentication mechanism to effectively mitigate the computational pressure of satellites. Zhu et al. [7] proposed a domain-based authentication scheme, in which each domain authenticates and manages the terminals locally to meet the concurrent authentication requests. However, most of the existing researches are aiming at the improvement of the performance of the algorithms, and not completely suitable for the architecture of the future SIN, even not fully utilize the advantages of the SDN architecture.

Scheme. Different from the traditional authentication technology, the satellite in the SIN can realize dynamic networking through SDN and get rid of the dependence on the ground gateway, thus the ground-to-satellite transmission delay can be greatly reduced. The authentication request of the terminal intends to join the network is received by the satellite and forwarded to the ground authentication center. The algorithm is lightweight enough, therefore, the authentication task can also be undertaken by the visible satellite. Distributed authentication provided by ground service center is also available [8]. Finally, with the help of the SDN flow table mechanism, the authentication result will be deliver to every forwarding device, including satellite node, and the terminal successfully join the SIN.

The entire scheme includes the authentication process, the de-authentication process, and the trusted communication process. The technical solutions in this part are shown in Fig. 4.

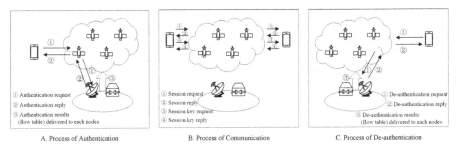

A. Process of Authentication   B. Process of Communication   C. Process of De-authentication

Fig. 4. Trusted authentication mechanism for networking entities

When the terminal intends to join the SIN, the terminal sends an authentication request to one of the satellite in the visible range, and the satellite forwards the request to the authentication center. The authentication center verifies the request (calculated by identifier *ID*, random number *R*, location *L*) and issues a digital certificate with the authentication center signature for the authenticated terminal as the network access license. When the terminals communicate with each other, they exchange their digital certificates and verify the identity of each other by verifying the signature of the authentication center in the digital certificate. A timestamp-based integrity verification is adopted to prevent attackers from tampering or replay attacks. When the communication ends, the de-authentication process is performed and the terminal access license is cancelled to prevent resources abusing and mitigate security risks.

Evaluation. The advantage of this solution is that it can make full use of the advantages of the SDN architecture and our proposed trusted authentication mechanism to realize the efficient networking of the networking entities and the efficient accessing of the terminals. Since the integrity verification and the digital signature mechanism are introduced to our scheme, the security of our scheme is guaranteed, and the replay attacks and the tampering attacks can be effectively resisted.

3.2 Dynamic Routing Optimization and Threats Isolation

Requirements. The constellation of the future SIN will become more and more complicated, hence the communication quality of the inter-satellite link will be complicated and time-varying as well. However, the proliferation of users and service demands will put forward higher requirements for the real-time routing optimization for SIN [9], security threats will also be severe due to the complexity of the network. Therefore, a dynamic routing optimization and threats isolation mechanism is highly demanded.

Related Work. Fei et al. [10] proposed a software defined network inter-satellite routing strategy, which involved traffic control mechanism, link cost calculation mechanism and congestion avoidance mechanism. Qi et al. [11] proposed a routing algorithm based on virtual node strategy and assigning different responsibilities to different satellites. In this scheme, the geostationary orbit satellite is responsible for routing calculation and the low-orbit satellite is responsible for routing and data forwarding. It is said that according to this scheme, the efficiency and reliability of the SIN routing can be both improved. However, most of the existing researches focus on the routing algorithm, and do not fully consider the time delay of the SIN. Besides, the routing calculation and its action mode will meet certain limitations when applied to the future resources-limited SIN environment.

Scheme. Different from the traditional terrestrial network, the future SIN will be multi-dimensional and its quality parameters will be time-varying, the processor processing power and the memory storage capacity of the satellite are limited more or less. Moreover, the on-board devices are difficult to recover from troubles, therefore any failure of the satellite should not affect the normal use of the entire network. The scheme we proposed can take advantage of the centralized management mode and the QoS (Quality of Service) routing mechanism, and finally provide a low congestion possibility and balanced routing path to meet the requirements of QoS [12].

Snapshot-Based Virtual Topology Routing Scheme. Due to the periodicity and predictability of the operation of the constellation network, the entire constellation operation period can be divided into several time slices. In each time slice, neither the connectivity nor the quality of the inter-satellite link will have major changes. Since the routing function of the network in the architecture is realized by the flow table delivered by the SDN controller, the flow table information can be calculated and distributed to each satellite node in advance. Therefore, the computing burden is greatly reduced and we realize the lightweight update of routing.

As is shown in Fig. 5. In different time slice, the connectivity of some inter-satellite links are changed.

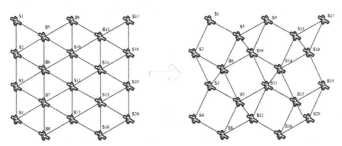

Fig. 5. The connectivity of the inter-satellite links changes according to the time slice switching

In the meantime, storing a large number of flow tables will consume too much storage space. By employing the snapshot mechanism, only the changed parts of the flow table will be stored. Therefore, the storage resources are also be saved.

Dynamic QoS Routing for SIN. This solution is used as an essential supplement to the above scheme. Due to the characteristics of the SIN, the quality of each routing path will change dynamically. The congestion of some links will lead to the degradation of communication quality, and the aggressive behaviors will threaten the privacy and the information security of users. Therefore, real-time optimization of routing is essential. We design a series of quality parameters to describe the quality of the inter-satellite communication links and set up corresponding thresholds for being triggered to optimize the routing. When the quality parameters is higher than the thresholds, the dynamic QoS routing is triggered to achieve dynamic optimization of the routing. Let the quality function of a path be:

$$quality(p) = \sum_{i=1}^{m} w_i^0 f_i(x)$$

Where p is the path number, $f_i(x)$ is one of the quality functions and w_i represents the weight of the i-th quality function. The link quality function covers multiple attributes such as delay, cost, bottleneck bandwidth, hop count of the path, packet loss rate and security coefficient etc. The normalization preprocessing of the values ensures that the output values of the quality functions are in the same range. At the same time, with the help of the security coefficients provided by the intelligent security management mechanism, the intra-range mapping of the security parameters is implemented, and the total security weight of the link is counted as:

$$Security(p) = \sum_{e \in p} Security(e)$$

When the dynamic QoS routing scheme is triggered, a weight vector is assigned to the quality function. For example, when the security of the communication is highly demanded by the system, the value of w can be set as (0, 0, 0, 0, 0, 1), which guarantee that the calculated routing is the safest in the current condition and the risky path will not be selected. However, when it is desired to balance the burdens among the links, the value can be assigned as (1, 0, 1, 0, 1, 0).

Evaluation. According to our experimental analysis, our scheme can take the advantages of the lightweight snap-based routing scheme and the real-time dynamic QoS routing scheme, ensuring that the management of routing is efficient and reliable.

3.3 Intelligent Security Management Mechanism

Requirements. The complexity of the future SIN architecture and the diversity of information services provided by SIN will expose it to more security risks [13, 14]. At the same time, due to the openness of the inter-satellite links and the massive number of the users, the malicious behaviors can be launched at any time, from any places. Therefore, the intrusion behaviors will be more difficult to predict. As a result, the intelligent anomaly detection capable for unknown attacks and security management strategies capable for mitigate the threats are demanded.

Related Work. Zhang et al. [13] proposed an intrusion detection model based on security domain partitioning, and proposed a distributed hierarchical collaborative intrusion detection scheme to ensure the security and reliability of satellite network. Wang et al. [15] proposed a security routing technique for spatial network based on intrusion detection, and designed a distributed intrusion detection model to provide a solution for secure spatial routing. However, the existing researches mainly focus on the detection algorithms and do not propose a systematic detection scheme or management strategies for SIN.

Scheme. Different from the traditional terrestrial network, the inter-satellite data transmission delay is longer and the satellite nodes are difficult to control, even worse, the failure of satellite nodes may cause irreparable consequence. Therefore, we propose an intelligent security management mechanism to realize the intelligent awareness of intrusion behaviors and comprehensive addressing of threats. In this way, the security and reliability of the network can be both improved.

Lightweight Intelligent Anomaly Detection. Considering that the resources of the Satellite is extremely limited, traditional anomaly detection methods based on artificial intelligence and big data analysis highly relay on the outstanding computing performance of the platform, and the model cannot be saved or loaded in a simple way. Therefore, we proposed a scheme in which the training module and the detection module are decoupled. The detection module can function without concrete platform, therefore, the resources consuming can be greatly mitigated.

We design an improved K-Means clustering algorithm based on Euclidean distance. By dynamically selecting the optimal cluster number (the value of K), the maximum discrimination between the positive and negative samples and the balance of overheads are realized. In addition, the clustering algorithm based on Euclidean distance is easy to realize and the trained model can be save and load in an understandable and easy-to-compute way. The algorithm and process is shown in Fig. 6.

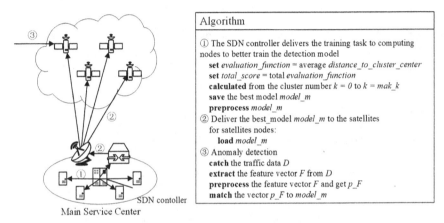

Algorithm
① The SDN controller delivers the training task to computing nodes to better train the detection model
set *evaluation function* = average *distance_to_cluster_center*
set *total_score* = total *evaluation function*
calculated from the cluster number $k = 0$ to $k = mak\_k$
save the best model *model_m*
preprocess *model_m*
② Deliver the best_model *model_m* to the satellites
for satellites nodes:
load *model_m*
③ Anomaly detection
catch the traffic data D
extract the feature vector F from D
preprocess the feature vector F and get p_F
match the vector p_F to *model_m*

Fig. 6. The lightweight intelligent anomaly detection algorithm and process

It is worth mentioning that the intrusion detection based on rules can also be employed to provide a protection with high accuracy and low false positive rates.

Threats Management Strategies. In the above-mentioned lightweight intelligent anomaly detection scheme, we realized the combination of the intelligent intrusion detection system and the rule-based intrusion detection system.

At the same time, due to the lightweight attributes of the intelligent anomaly detection system and the limited resources of the satellite in our scheme, the detection and treatment released by single node is likely to have a poor performance. Along with the time-varying relative position between satellite and ground, repeated detection towards anomaly behaviors may lead to unnecessary resources overhead. Therefore, with the help of SDN architecture, the detection results are transmitted to the SDN controller in real time, and the traffic is scheduled to the ground security module for safer treatment according to the security strategies. Then the SDN controller analyzes the results conveyed by satellite nodes, and then transmit the analyzed results and manage the entire network with the help of flow table.

Combined with the dynamic routing optimization and threat isolation scheme, we achieve the goal that the network capabilities and security are improved (Fig. 7).

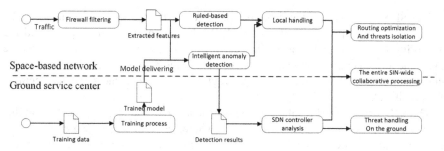

Fig. 7. Security protection strategy based on collaborative processing of the ground service center and the space-based network

Evaluation. Through experimental analysis, we validated and tested the algorithm by using the distributed computing platform, and evaluated it with the KDD CUP 99 data set. Combined with the rule-based intrusion detection system, our method ensures the awareness of unknown attack behaviors and accurate detection for known attacks. This will guarantee the real-time detection and effectively avoid the spread of threats caused by the long time delay if the detection tasks are assigned to ground service center. Besides, the SDN architecture provide the comprehensive protection capabilities for the entire framework (Table 1).

Table 1. Part of the experiment results perform by our training process. (Due to the best model model_m consist of over one hundred clusters, we select part of them as the display)

Cluster number	1	2	3	4	5	6
Label	smurf	normal	neptune	normal	neptune	portsweep
Accurate count	227020	15103	263	2479	4298	73
False count	0	1235	17	1	0	1
Accurate rate	100%	92.44%	93.92%	99.96%	100%	98.65%
Cluster number	7	8	9	10	11	12
Label	ipsweep	normal	smurf	normal	neptune	normal
Accurate count	1133	4340	66	5137	354	34
False count	188	44	0	6	9	3
Accurate rate	85.77%	99.00%	100%	99.88%	97.54	91.89

4   Conclusion and Prospect

With the development of SIN-related technologies in various countries and the increasing demands for space information services, it is imperative to build a highly available and reliable SIN. However, in the face of limited space resources and highly complex network environment, traditional network management methods will be difficult to adapt to future SIN.

The SDN-based security architecture proposed in this paper greatly enhances the security and availability of the future SIN. By decoupling the control plane and the forwarding plane, we realize a flexible network management mode. In this way, we realize the unified configuration and management of the network, the authentication for networking entities, the optimization of the routing, and the awareness and management of anomaly behaviors.

Through experimental results and theoretical analysis, the proposed scheme in this paper can achieve efficient management of the whole network. Under the premise of saving computing resources and storage resources as much as possible, the scheme greatly improves the security performance and service capabilities, and has great reference value for the future construction of the future SIN.

In the context in which all countries in the world have proposed SIN development plan, the architecture of the future SIN will be highly complex and the demands for services will be diversified. In the future, we will continue to study on security mechanism designed for SIN to provide the network more secure protection. Besides, we will research on the more optimized service composition algorithm and more secure authentication algorithm to provide more secure and available services.

References

1. Li, D., Shen, X., Gong, J., et al.: On construction of china's space information network. Geomatics Inf. Sci. Wuhan Univ. **40**, 711–715 (2015)
2. Li, F., Yin, L., Wu, W., et al.: Research status and development trends of security assurance for space-ground integration information network. J. Commun. **37**(11), 156–168 (2016). https://doi.org/10.11959/j.issn.1000-436x.2016229
3. Sezer, S., Scott-Hayward, S., Chouhan, P., et al.: Are we ready for SDN? Implementation challenges for software-defined networks. IEEE Commun. Mag. **51**(7), 36–43 (2013)
4. Wang, T., Chen, H., Cheng, G., et al.: Research on software-defined network and the security defense technology. J. Commun. **38**(11), 133–160 (2017). https://doi.org/10.11959/j.issn.1000-436x.2017221
5. Zhang, C., Cui, Y., Tang, H., et al.: State-of-the-art survey on software-defined networking (SDN). J. Softw. **26**(1), 62–81 (2015). https://doi.org/10.13328/j.cnki.jos.004701
6. Zhu, H., Wu, H., Zhao, H., et al.: Efficient authentication scheme for double-layer satellite network. J. Commun. **40**(3), 1–9 (2019). https://doi.org/10.11959/j.issn.1000-436x.2019058
7. Zhu, L., Wang, L., Li, J., et al.: New entity authentication and access control scheme in satellite communication network. J. Commun. **39**(6), 73–80 (2018). https://doi.org/10.11959/j.issn.1000-436x.2018103
8. Zhang, Z., Zhou, Q., Zhang, C.: New low-earth orbit satellites authentication and group key agreement protocol. J. Commun. **39**(6), 146–154 (2018). https://doi.org/10.11959/j.issn.1000-436x.2018102
9. Ma, J., Qi, X., Chen, C.: Routing algorithm based on congestion avoidance in satellite networks. J. Jilin Univ. (Sci. Ed.) **57**(2), 357–362 (2019)
10. Fei, C., Zhao, B., Yu, W., et al.: A routing strategy for software defined satellite networks considering control traffic. J. Beijing Univ. Aeronaut. Astronaut. **44**(12), 2575–2585 (2018). https://doi.org/10.13700/j.bh.1001-5965.2018.0343
11. Qi, X., Ma, J., Liu, L., et al.: Routing optimization based on topology control in satellite network. J. Commun. **39**(2), 11–20 (2018). https://doi.org/10.11959/j.issn.1000-436x.2018020

12. Jiao, Y., Tian, F., Shi, S., et al.: Multipath routing based congestion control strategy for LEO satellite networks. Electron. Des. Eng. **26**(18), 119–123+128 (2018)
13. Zhang, W., Xu, Y.: Research on the intrusion detection model based on the division of the security field in satellites network. Trans. Shenyang Ligong Univ. (6), 9–13 (2008)
14. Niyaz, Q., Sun, W., Javaid, A.: A deep learning based DDoS detection system in software-defined networking (SDN). ICST Trans. Secur. Saf. **4**. (2016). https://doi.org/10.4108/eai.28-12-2017.153515
15. Wang, X., Zhang, Z., Li, M., et al.: A secure routing technology based on intrusion detection in space information network. Appl. Electron. Tech. **41**(4), 101–104 (2015). https://doi.org/10.16157/j.issn.0258-7998.2015.04.024

Research on Intelligent Task Management and Control Mode of Space Information Networks Based on Big-Data Driven

Xiaogang Yu and Qi Wang[(⊠)]

Beijing Institute of Remote Sensing Information, Beijing, China
wq960121@outlook.com

Abstract. The space information networks provide a rich space, time, frequency spectrum resources, meet all kinds of scene mission requirements, especially the rapid development of information technology and the interaction of human life fusion, and the global data presents the characteristics of explosive growth and massive convergence, artificial intelligence has advantages such as flexibility, adaptability and low robustness in the direction of information fusion. On the basis of studying the framework of space-based information network, an integrate task management and control mode based on big-data driven space-based information networks and Internet of things is proposed. Artificial intelligence technology is used to solve the problem of the front-end requirements of task management and control, and the ratio of resource utilization to actual profit of joint information network load points is improved, at the same time, it lays a foundation for realizing autonomous task planning.

Keywords: Big data · Artificial intelligence · Space & Earth information networks · Task control

1 Introduction

The information explosion in the era of big data will have a significant impact on economic development, military construction, social order, people's life and other directions. Artificial intelligence has become another subversive change in the field of information technology after cloud computing, Internet of things and big data. With the in-depth development of artificial intelligence in various fields, breakthroughs have been made in speech recognition, human-computer game and unmanned driving built by artificial intelligence technology, which makes artificial intelligence have a huge prospect in the field of structured information application [1]. Under the background of the global high-speed information transmission technology is about to achieve comprehensive cloud services, and the information industry system ushers in great changes, "everything connected, everything connected" has derived the urgent demand of integrated information network and AI technology in deep IOT applications. Under the ideal condition of whole network sampling, the data samples of human activities are generated all the time [2]. In essence, these data generated by human activities are collectively referred to as big data,

© Springer Nature Singapore Pte Ltd. 2020
Q. Yu (Ed.): SINC 2019, CCIS 1169, pp. 112–131, 2020.
https://doi.org/10.1007/978-981-15-3442-3_10

and the abstract representation is language, optical image, audio and text. As for the measurement of information entropy, the measurement of information itself is certain, but the value of information is different from the demand of information receptor It is also different from the response of the receptor to the information situation, which depends on the position of the information receptor in the human social structure. Often we pay more attention to the information receptor with higher positive feedback to the information response. Therefore, ideally, we hope to realize the space-oriented information network and the ground network, and combine big data mining and artificial intelligence technology to make the core users independent Push information forecast and continuous state make spatial information network become a front-end tool of "managed by people, unattended" instead of being processed object, and shift more attention to the application level of user demand and task response processing.

2 Information Source Characteristics and Engineering Analysis of Spatial Information Network Task Management and Control

Aiming at the three kinds of data of image (spectrum, microwave), audio and text contained in the open source information of the ground database, the network model is established by the deep learning technology such as CNN convolution neural network [4], deep convolution reverse graph network, RNN cyclic neural network, LSTM memory network [4], the network model is based on CNN image recognition network model, and the text classification cyclic network model is based on PTB data set For the core user's needs, divide the message storage pool by message type in the open source database, parallel the distribution of global key objects in other source databases, urban traffic monitoring data, geographic resource prediction data, real-time target access data, navigation and positioning data, military strike evaluation data, etc., and conduct the calculation according to the prior probability of D-S evidence theory [5] and network model Information fusion, the analysis of events in complex and huge source data, forms the task requirements driving the space-based platform, and realizes the closed-loop link of "big data - demand mining - spatial network node - processing - feedback - demand mining correction". Because neural network provides a kind of mapping "dark box", so using network engineering models such as LSTM, VEA, FFNN [6] can better solve the problem of feature information quantification of multi-type source data (image, text, voice, etc.). In deep learning, LSTM network or NTM neural Turing machine can be used to learn complex sequences [7], CNN network can be used to learn image information, in addition, On the other hand, Networks [4] such as VAE and SAE can be used to design adaptive classification algorithm and build powerful association analysis model.

The resource space for big data analysis is characterized by wide coverage, large redundancy, complex information and great security threat. The traditional serial access mechanism seriously affects the timeliness of data processing. Distributed data access and management provide reliable and efficient solutions for the application of big data. Through intelligent cloud computing, i.e. distributed data access point, the data can be processed in parallel and adjusted by the intelligent cloud of the center coordinates and initiates tasks. The distributed end of LAN cloud computing combined with source sensor can provide good data stability and timeliness support for task management and control of spatial information network.

3   Analysis of Space-Based Mission Control Situation

At present, in the space-based operation system, the operation process of each space node is a closed-loop link of "demand control satellite processing user", which mainly includes demand acceptance and planning, task control, satellite measurement and control, load action, data receiving and transmission, information processing, calibration processing, data production and other links. Among them, the demand acceptance and planning layer accepts the target access, geological survey, positioning and navigation needs of all kinds of users at all levels, carries out classified processing and independent deployment, eliminates demand conflicts, defines the requirements for load task implementation, and submits them to the task management and control subsystem for review and Implementation; in addition to the normal acceptance needs, it is always faced with the emergency adjustment of space-based tasks, and the whole process needs to be adjusted In addition, a large number of human and material resources have been invested. In addition, the measurement and control window is limited, the tracking and receiving resources are limited, etc. as the front-end demand acceptance of the whole space-based operation system, the information analysis ability tends to be insufficient, such as the global situation, and the contradiction between the current satellite management and control ability, making the ratio of satellite tasks to capabilities and life at a low level, the business value effectiveness and task resource consumption The consumption ratio is also at a relatively low level, and the lack of demand analysis capability is gradually amplified in the serial process, which leads to a great discount in the timeliness and efficiency of the overall operation of the space-based control system.

4   Spatial Information Network Driven by Multi-source Big Data

After the rise of big data wave, after years of practice, the development of big data on the Internet and mobile Internet has been relatively mature, and big data has moved from concept to practice. However, the space-based task management and control system and the processing and utilization of space to big data still use the traditional independent operation system, which has a single mode and can not make changes in time according to the changes and needs of scenarios and users, Means that the importance of data-driven spatial information network for "intelligent decision-making and management" is appreciated. Through efficient data analysis, the output of spatial information network task management and control system is refined for operation and control, the production form is optimized, the operation and control mode is adjusted, and the efficiency conversion rate is improved. Remote sensing data, earth environment data, astronomy data, meteorology data and space environment data account for a large proportion in the space-based big data, and the situation function dimension of events is often high. As far as the military target reconnaissance demand of satellites is concerned, how to accurately provide non cooperative moving target position information to the satellite nodes in the space network is the target trend information number such as remote sensing data, According to the fusion and output of prior conditions such as satellite access arc section and meteorological data, not a single factor can decide; secondly, strengthening the emergency management of geological disasters still requires the real-time sampling data of ground systems such as remote sensing data, earth mapping data, meteorological data and geological data provided by various platforms of spatial information network, which come from different directions and huge data, The response ability of spatial information network tests the load-bearing ability of spatial information network. How to change passive into active, relieve the pressure of demand response, build "intelligent operation and control" of spatial information network, and improve the comprehensive ability of operation and control of spatial information network should be studied from the structure of spatial network and ground system and the level of big data drive.

5 Space & Earth Integrated Information Networks

The space information network management and control system integrates the ground operation management network (traffic monitoring sub network, driverless vehicle operation sub network, UAV operation sub network, etc.) into the framework of the satellite ground integrated information service operation system based on the structured space-based information management and control network [8], with the space management and control center and the ground management and control center as the core of the system's centralized management and control, and the satellite ground big data as the support In the data service layer, independent task demand decision-making model is established as the driving engine of the joint information management and control framework to inject fuel into the task planning layer and provide normalized information services for the user layer.

The sky earth integrated information network management and control system has the ability to accurately sense the position of network members and access the orbit calculation. It supports the following couple access and rapid chain building of network members, space vehicles, ground vehicles and other terminal nodes. It has the

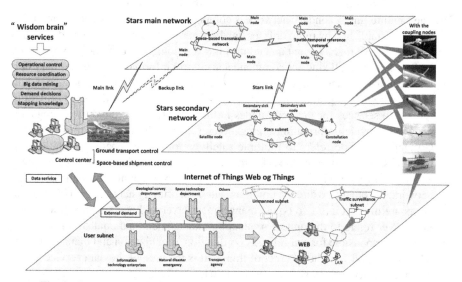

Fig. 1. Space & Earth information networks "Intelligent operation" control mode

characteristics of safe, reliable, controllable and manageable operation, and the operation of "network without center, information with center". It is an important support of the capacity of the satellite earth joint information network system Bracing. As shown in Fig. 1, the joint network consists of space-based control system and ground-based control system.

The space-based management and control system includes satellite link primary network [8, 9], satellite link secondary network [8, 9], and space-based operation and control system; the ground-based management and control system includes ground-based operation Internet of things (user subnet, unmanned transportation system subnet, traffic monitoring system subnet, etc.), and ground-based operation and control system; the user layer includes: national defense related departments, national geological survey related departments, national natural disaster prevention and control system Emergency related departments, transportation departments, space technology test related departments, information technology group industries, satellites, space shuttles, near space platforms, UAVs, etc., with user level users as the final service objects.

(1) The satellite link primary network is composed of high orbit communication backbone network and space-time reference network, including space-based transmission network, space-time reference network and high orbit satellite established through radio or laser link between satellites. The satellite link primary network is not only the backbone convergence network of the whole space-based network monitoring, measurement and control, navigation, timing and other management and control information, but also the space-based information network inter satellite management and control system Command and dispatch center of. Each backbone node is not only a single satellite with certain computing and storage capacity, but also a cluster/ constellation supporting high-performance computing and storage. The differential setting of backbone nodes can effectively meet the classification of spatial information and maximize the use of spatial resources.

(2) The satellite link secondary network is composed of satellite link subnet and LEO satellite established by radio or laser link between satellites. Ordinary nodes generate and receive management and control information. The secondary convergence node coordinates all kinds of command and control information of the primary network, shares management and control information in the satellite link subnet, and interfaces with the primary network. As shown in Fig. 1, in addition to providing relay data services for the main nodes in the network, the secondary aggregation node also provides information aggregation services for the user service layer nodes. Constellation node refers to multiple satellites forming cooperative relationship through satellite building link. There are two ways to generate or receive management and control information: one is to gather several points through secondary network, the other is to directly communicate with primary network backbone node.

(3) Users are divided into two types: management layer and business layer. Management layer refers to the unit of production management and control information, which is responsible for centralized coordination, organization and management of all kinds of nodes in the Internet of things. It is the core of the data service of the information network of the integration of Space & Earth. For example, receiving the

navigation service request of the ground unmanned transportation system and coordinating the stratospheric nodes to obtain meteorological and other environmental resources, planning the response tasks of optical satellite network and space-time reference network, and returning the task data to the ground unmanned transportation system. The business layer refers to the mobile vehicles that need to access the space-based or ground-based IOT network management and control system in a short time based on the task request of the management layer, including satellites, space shuttles, adjacent space platforms, UAVs, unmanned vehicles, ships, etc. According to the different requirements of task security level and time sensitivity, the satellite link primary network and secondary network can be dynamically linked as required.

(4) The main responsibilities of the space-based operation control system include managing the registration information of satellites in orbit, monitoring the status of space-based network in real time, planning node tasks according to the task uplink instructions of the ground control center, coordinating and scheduling the resources of each node, and regularly feeding back the operation status of space-based network and the health status of all kinds of satellites to the ground.

(5) The ground-based operation control system has the backup of the space-based operation control system. Its main responsibilities include accepting or producing the task requirements of the user management, managing all kinds of ground operation system nodes, real-time monitoring the ground network node status, and acting as the space-based temporary emergency management platform when the space-based operation control system fails. The integrated management and control center relies on the Internet plus the massive data of the Internet, and through the means of artificial intelligence, excavate and analyze the information about natural disasters, national defense situations, hot events and other related information, and transform them into regularized languages, and create business requirement uplink space management and control centers and all kinds of operation subsystems of the Internet of things, thus driving the ground monitoring system, the ground navigation operation system and the near space. The operation of service chains such as platforms and satellites provides users with spatiotemporal continuous service data, while the ground system has the ability to directly respond to users' needs in real time.

6 Technical Requirements of Intelligent Task Management and Control in the Integration of Space & Earth Networks

It can be seen that in the collaborative application layer of the Space & Earth integrated information network, the conventional management and control mode that requires a lot of human-computer interaction has been unable to support the huge network system. As the "heart" of the satellite earth joint information network system, the research on the intelligent centralized management and control technology is of great significance, and the business demand as the front-end is the space-based or ground-based network efficiency Giving full play to the source power, autonomous task demand decision-making technology plays an important role.

In order to realize the centralized autonomous task demand decision-making under the background of the information fusion between the Space & Earth, the following key technologies are needed: big data distributed management technology [1]; PTB data set language model based on the cyclic neural network; speech model based on BIRNN neural network [7]; pattern recognition technology based on the dense net convolution neural network [10]; information fusion interpretation and analysis technology; multi-source information network interaction Data security management policy. Among them, the big data distributed mining management integrates the Internet and the internal operation data network to take the initiative in data crawling, and selects the training set that has been classified from the data. In this training set, the data mining classification technology is used to establish the classification model to classify the unclassified data. Compared with BP feed-forward network, RNN is a feed-forward neural network with time connection. It has state and time connection between channels, which means that the order of input will determine the output, which is widely used in text understanding, speech recognition and other fields. Therefore, it focuses on the optimization algorithm model of LSTM long and short-term memory network in RNN to realize Based on information source big data mining, all kinds of event situation information and feedback concerned by users are formed and event description key value pairs are inserted into the running event queue of integrated system. Convolution neural network and convolution neural network, which also have high recognition degree and generalization ability in the field of digital image processing, can quickly segment and recognize images, obtain regular description information of key objects of images, and form event description key value pairs to insert into the operation event queue of integrated system. Under the background of complex information sources and huge amount of data, how to carry out information association analysis needs to conduct in-depth research on information fusion and other related theories, carry out association analysis on the acquired event files to form a comprehensive decision-making meta task, and the generated meta task directly enters the operation driving link. Because the multi-source platform is used as the input environment of meta task output at the same time, the information is not safe How to ensure data security under the background of data security is the need for policy and technical research.

At present, CNN and RNN are in the core application position in the field of deep learning digital image processing. RNN recurrent neural network is a feedforward network with time connection. The input information of neuron includes not only the output of the former neural cell layer but also its own state in the previous channel. Because most of the data do not have the change of time line in form, they can In the form of a sequence. A picture, a paragraph of text, a paragraph of speech can be input by a pixel, a paragraph of text or a voxel. Therefore, the time-related weight and the forgetting related weight describe what happened in the previous step of the sequence, so as to predict what will happen in the next step. It has been proved that LSTM can be used to learn and understand the complex sequence [7], but at the same time, it will occupy a large amount of Computing resources, so we can use the network to filter out the description we are concerned about in the data layer.

D-S evidence theory was first proposed by Dempster in 1967 and further developed by his student Shafer in 1976. It belongs to the category of artificial intelligence. It was first applied to expert system and has the ability to process uncertain information. It is usually used for fault diagnosis and system running state detection. As a kind of uncertain reasoning method, the main characteristics of evidence theory are: it satisfies the weaker conditions than Bayesian probability theory; it has the ability to express "uncertain" and "unknown" directly. From the trust degree of proposition, namely the basic probability distribution and combination rules, it can calculate the new basic probability distribution function of the fusion information produced by the common action of each topic, and The probability distribution can be calculated by neural network, so in the later stage of information fusion, the fusion decision results can be given by combining neural network or as a means of confidence analysis [5, 11].

7   An Autonomous Decision Technology of Information Fusion Task Demand Based on Neural Network

Imitating the three main channels of human brain to obtain information: language, voice, image (spectrum) and the thinking process of the actor to understand the behavior of fusion information source data, making information fusion decision according to the mathematical principles of machine learning and information fusion, this paper proposes an information fusion decision-making mode based on the artificial neural network technology and D-S evidence trigger output (Fig. 2):

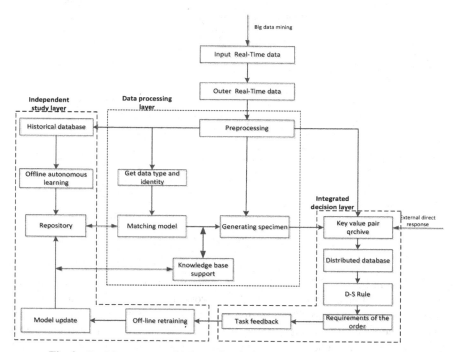

Fig. 2. Decision process of information fusion based on artificial intelligence

As shown in Fig. 3, the decision-making body of information fusion is divided into real-time data processing layer, offline autonomous learning layer and fusion decision-making layer. The real-time data processing layer will slice, identify and classify the feature data of users' concern from the shallow scan of big data in the future, generate the basic event samples marked with natural distribution probability, component elements and time-space stamps, and process the event samples in the fusion decision-making layer Data association analysis establishes event files, generates event analysis reports and corresponding response tasks by information fusion, plans production control information of decision-making level and transmits it to all levels of IOT sub networks. In this process, different kinds of data need to be spatiotemporal correlated and identified, and the corresponding mathematical model needs to be constructed, which integrates the memory neural network and convolution network. In this paper, a kind of coding memory propagation neural network model with fusion recognition ability and a fusion decision model based on event samples are proposed.

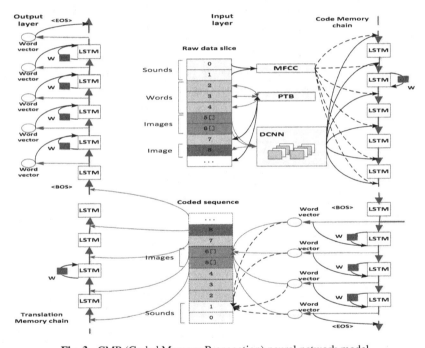

Fig. 3. CMP (Coded Memory Propagation) neural network model

As shown in Fig. 4, CMP neural network slices the original source data according to text, image and voice by category in the network input layer to generate the basic elements of the voxel vector, word vector, continuous frame pixel bit map with spatiotemporal correlation and single frame pixel bit map. The encoding memory chain (LSTM) recognizes and encodes the pixels and the acoustic blocks in the original data slice and retains the temporal and spatial characteristics of the original data. After encoding memory chain, the original data will be described as encoding sequence in a unified way. The translation memory chain (LSTM) selectively forgets and retains the encoding sequence units and outputs the processing results in the output layer. In a word, CMP network can recognize multi-source complex sequences and retain the temporal and spatial characteristics of the sequences. For a frame of information source data processing, event samples are generated through CMP network first. Event samples are classified into files according to event time-space stamp and priority, and respond to user level's direct and clear task requirements synchronously (such as SAR strip imaging in designated target area or traffic image, weather and ground navigation in a certain period of time in designated target area). Because of the direct demand information from users It is clear that it can be directly processed to archive event samples, as shown in Fig. 5. For an observation object, two pairs of event samples with spatiotemporal correlation are matched according to formula (1). The matching rule is: when the strength of the correlation key ΔP_{AB} meets a certain limit, the event samples will enter the complete collection waiting sequence. Among them, function F($\cdot$) is the response function of two event correlation weight, and specific algorithm needs to be designed. Parameter P_i is the natural distribution probability of an event, function H($\cdot$) represents the similarity of an observation object in two events, and parameter O_i reflects the value of a specific observation object in a specified event. The function D($\cdot$) represents the Euclidean geometric distance of time and space between two events, ΔT_{AB} and ΔR_{AB} are the time and space intercept of two events.

$$\Delta P_{AB} = F(P_A, P_B, H(O_A, O_B), D(\Delta T_{AB}, \Delta R_{AB})) \tag{1}$$

Once the sample set satisfies the D-S symptom set of the object, the meta task of fusion decision-making is performed. According to D-S evidence theory, let $\{\Theta\}$ be a recognition framework, or hypothesis space. There are:

(1) Basic probability assignment (BPA for short). BPA on recognition framework $\{\Theta\}$ is a function M($\cdot$) of $2^{\Theta} \rightarrow [0, 1]$, which is called Mass function, and meets the following requirements:

$$\sum_{A \subseteq \Theta}^{n} m(A) = 1, m(\varnothing) = 0$$

Among them, A^{Θ} which makes m (A) > 0 is called focal elements.

2) Trust function, also known as Belief function, is defined on the recognition framework $\{\Theta\}$ based on BPA m as follows:

$$Bel(A) = \sum_{A \subseteq \Theta}^{n} m(B)$$

3) Likelihood function, also known as Likelihood function, is defined based on BPA m in the recognition framework $\{\Theta\}$:

$$Pl(A) = \sum_{B \cap A = \varnothing}^{n} m(B)$$

In the evidence theory, for a certain hypothesis a in the recognition framework $\{\Theta\}$, according to the basic probability assignment BPA, the trust function bel (A) and the likelihood function Pl (A) are respectively calculated to form the trust interval

[Bel (A), Pl (A)], which is used to express the confirmation process of a certain hypothesis Dempster's compositional rule, also known as evidence compositional formula, is defined as follows, for $\forall A \subseteq \Theta$, the Dempster compositional rule of two mass functions M1 and M2 on $\{\Theta\}$ is as follows:

$$m_1 \oplus m_2(A) = \frac{1}{K} \sum_{B \cap C = A}^{n} m_1(B)m_2(C)$$

$$k = \sum_{A_i \cap B_i \cap C_i \ldots = \phi} (m_1(A_1)m_2(B_2)m_3(C_3) \ldots m_i(N_i))$$

Where k is the normalization constant. The final output generally follows the following rules [5]:

(a) The determined event pattern should have the maximum reliability function value;
(b) The difference between the reliability function value of the event mode and other modes determined is greater than a certain inner limit;
(c) The value of uncertain mode function must be less than an inner limit;
(d) The value of reliability function of decision mode must be greater than the value of uncertainty reliability function.

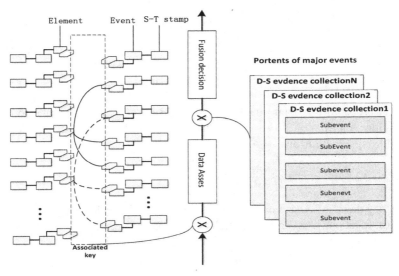

Fig. 4. Event samples and fusion decision model

With the increase of data volume, the matching rules of event files (structured data) need to be further optimized. At the same time, considering the use of distributed processing framework can further optimize the timeliness of data processing in the fusion layer.

In the autonomous learning layer, CMP network and D-S symptom set will update the link parameters of each layer of the network adaptively according to the results of real-time data processing and demand response, remove the real-time data that does not meet the decision rules, and retain the original data that meets the decision rules as the model learning samples to insert into the target recognition database. The data processing model, especially the generalization of neural network model, the ability of feature recognition and the comprehensive improvement of fault tolerance rate are realized through the off-line learning layer forward feedback. In the following, we will focus on how to generate "intelligence" reasoning and requirements based on the structured data of dynamic big data mining of Space & Earth integrated information network, as well as task generation and feedback iteration of heterogeneous computing nodes of Space & Earth integrated information network.

In the process of dynamic acquisition of complex information, in addition to obtaining real-time high-value information, how to integrate the information between the information, obtain the rules of event situation evolution, locate the key points of induced events, form the demand for core target monitoring, and become the key of intelligent demand decision-making technology research.

At present, there are some difficult problems in the research of event intelligent reasoning, among which the most important one is that the event knowledge association is often very complex. For the same objective event representation, there are many possible stacking factors, even involving many different behavioral representations. For example, taking the early warning of conflict situation among countries as an example, there are many factors that lead to the conflict between countries, such as trade competition, even subjective politics, religious accumulation, etc. taking the armed conflict between countries A and B as an example, the conflict is generally It is an explosive high-dimensional nonlinear mapping with certain factors superimposed over time, which has the contradiction of randomness and certainty. Because of the different experience and knowledge of experts in different fields, their judgment of induced conflict is only based on their own experience and knowledge, but not fully considering the experience and knowledge of experts in other fields. In this case, the judgments given by experts in different fields are often different or even contradictory, which leads to the inability to accurately locate key elements.

In order to solve the above problems, it is necessary to fuse the information of different event symptom domains corresponding to the same situation. D-S evidence theory has been widely used in target recognition, decision analysis, fault diagnosis, condition monitoring and other fields because of its strong ability to deal with uncertain information.

D-S evidence theory requires that the basic probability distribution is known when synthesizing evidence, but in intelligence reasoning, the distribution between different symptom domains of events is generally only based on experience, so it is difficult to determine the distribution probability, which limits the practical application of the theory. In this paper, BP neural network and D-S evidence theory are combined to make full use of BP neural network's self-learning, self-adaptive and fault-tolerant ability and apply them to the process of determining the reliability mapping of different symptom domains. An intelligence reasoning model based on BP neural network and D-S evidence theory is established to more effectively express and process uncertain information, so as to reasonably synthesize uncertain information, The experience and knowledge of the expert system in the same field enable the end information fusion reasoning system to automatically complete the transformation from information to requirements. Figure 5 is the schematic diagram of D-S fusion process:

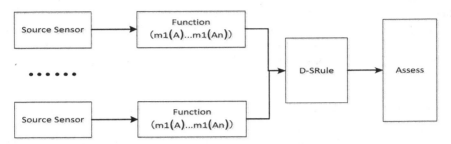

Fig. 5. Structured data and fusion decision model

In the intelligence knowledge reasoning system, BP neural network and D-S evidence theory are introduced. BP neural network is used to obtain the basic probability distribution of different event symptoms. According to this probability distribution, D-S evidence theory is used to fuse the information of multiple symptoms of event evidence body to obtain the reasoning results. Firstly, according to the event database of dynamic analysis, all symptoms corresponding to each evidence are obtained by feature extraction. In order to construct the basic probability distribution function of all symptoms to each event, the acquisition process is usually complex and non-linear. From the characteristics of BP network, as long as the network has strong generalization ability after learning a large number of samples, it can better describe the complex non-linear mapping relationship, thus determine the basic probability distribution, and establish the evidence body pair as the observer, The confidence mapping function of the event.

Now, the simple structured data of a certain identification sub event established after multi-source big data mining through CMP network is simulated as follows (Table 1):

Table 1. Sub event structured data

Event label	Score	Description
Event	0.92	"The Country(A) side has main intentions with Country(B)"
Tag	0.05	Bad politics
Time	1.00	August 10, 2020
Hotspot	1.00	Island, Missile, Economic blockade, Warship
Position	1.00	Island

After the intelligence knowledge is structured, it often needs deep information mining to obtain deeper intelligence knowledge. Therefore, after the rapid classification and recognition of intelligence, the score evaluation is used to measure the information value of this kind of intelligence. Take the messages of politics class as an example to construct the text data sample set as follows (Table 2):

Table 2. Deep evaluation sample space (part)

Politics score	Sample (Description)
0.9	"The two countries have close diplomacy"
0.8	"The two countries have maintained diplomacy"
0.7	"The two countries have a relatively close diplomatic relationship"
0.6	"Normal diplomacy between the two countries"
0.3	"Diplomatic tensions between the two countries are high"
0.1	"Relations between the two countries have broken down"

Among them, the mapping of evaluation of "politics" is quantified as the weight of interval [0, 1]. Now we define that the closer and easier the "political relationship" is, the closer the evaluation weight is to 1. On the contrary, the more tense the political and diplomatic atmosphere between the two countries is, the lower the corresponding description score is. The natural language of the sample space has the characteristics of different scores. The sample set will be trained by RNN neural network to get this kind of High dimensional feature mapping from natural language description to information evaluation score.

Now, we simulate multiple groups of sub event structured data (tags). The background is set as follows: according to certain structured data, we can infer the main factors and key spatiotemporal information that induce the armed confrontation between A and B, and trigger the monitoring requirements of spatiotemporal observation or "intelligence" tracking of heterogeneous platforms of the sky earth integrated network. In order to facilitate the experiment, the label description and reasoning structure are simplified. Three kinds of test labels, namely "political relation", "military relation" and "economic relation", are designed according to the space-time distance and key words matching (Tables 3, 4 and 5).

Table 3. Politics (Label_1)

Event label	Score	Description
Event	1.00	"A renounce diplomacy on B"
Tag	0.23	Politics
Time	1.00	August 10, 2020
Hotspot	1.00	"A{Somebody(import) Spoke}", "B{The Government Policy}"

Well, it is assumed that the evidences inducing the armed conflict between countries A and B are: "diplomatic relations"; "trade relations". Now, it is assumed that "diplomatic relations" are determined by three kinds of symptoms: political events, economic events and military events. Now, it is assumed that diplomatic relations between countries A and B are an independent scoring system, only considering three kinds of factors: political, economic and military. Trade between countries A and B As an independent scoring system, relationship only considers political, economic and military factors.

Table 4. Economics (Label_2)

Event label	Score	Description
Event	1.00	"A imposed an economic blockade on B"
Tag	0.16	Economics
Time	1.00	August 10, 2020
Hotspot	1.00	"A", "B", "economic sanctions"

Table 5. Military (Label_3)

Event label	Score	Description
Event	1.00	"A get conflict on B"
Tag	0.02	Military
Time	1.00	August 10, 2020
Hotspot	1.00	"A", "B", "Destroyer(A)", "Destroyer(B)"

Now, the BP neural network with input layer of 3, double hidden layer of 3 and output layer of 1 is used to simulate the evaluation mapping of diplomatic relations between countries A and B, which is also used as the evaluation mapping of trade relations. The threshold of political, military and economic scores is set as [0, 1]. The higher the score is, the healthier the corresponding relationship is. On the contrary, the lower the score is, the more hostile the corresponding relationship is, as shown in Tables 6 and 7.

Table 6. Score of foreign relations evaluation system

Label	Score	Description
Politics	0.91	Good
Economics	0.43	Not good
Military	0.02	Bad

And the corresponding reasoning structure is (Fig. 6):

The training data set of double-layer feedforward neural network is constructed. The training measures "diplomatic relations" and "trade relations" from three dimensions, including political, economic and military perspectives. The structure of BP network is as follows (Fig. 7):

To construct the training sample data set of diplomatic relations and trade relations between countries A and B and their political, military and economic mapping, it can be seen from the data that the two types of mapping relations are different, as shown in Tables 8 and 9:

Table 7. Score of trade relations evaluation system

Label	Score	Description
Politics	0.63	Good
Economics	0.91	Good
Military	0.81	Good

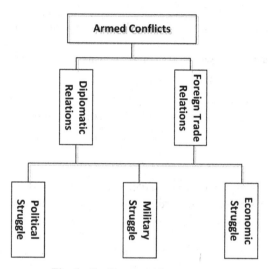

Fig. 6. Conflict reasoning structure

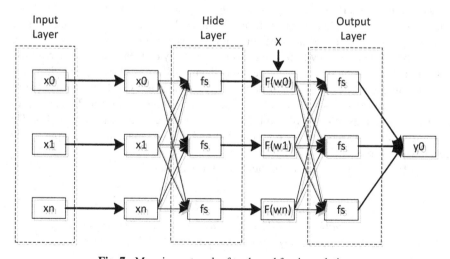

Fig. 7. Mapping network of trade and foreign relations

Table 8. Symptom set of diplomatic relations (part)

Name	Sample_0	Sample_1	Sample_2	Sample_3
Politics	0.21	0.21	0.87	0.91
Economics	0.13	0.36	0.98	0.92
Military	0.02	0.11	0.91	0.89
Label	0.0	0.0	1.0	1.0

Table 9. Sample set of trade relationship symptoms (part)

Name	Sample_0	Sample_1	Sample_2	Sample_3
Politics	0.61	0.51	0.37 .	0.91
Economics	0.13	0.36	0.98	0.92
Military	0.85	0.13	0.91	0.19
Label	0.0	0.0	1.0	1.0

Among them, BP neural network is used to obtain the basic probability distribution of different relationship symptoms. According to this probability distribution, D-S evidence theory is used to fuse multiple symptoms of the two types of relationship, so as to obtain the reasoning results.

First, according to the real-time update of the event tag library information data, through feature extraction, we get all the symptoms corresponding to each kind of mutual relationship. To construct the basic probability distribution function of all symptoms for each event, the acquisition process is usually complex and nonlinear. From the characteristics of BP network, it can be seen that as long as the network has strong generalization ability after learning a large number of samples, it can better describe this complex nonlinear mapping relationship, and thus determine the basic probability distribution. Then we use D-S evidence theory to fuse the basic probability distribution of each symptom and get the final fusion result. The schematic diagram above is shown in Fig. 8.

According to the designed map structure, DS evidence theory and BP neural network, training and data analysis are carried out. The scores of {label_1, label_2, label_3} are taken as the input layer tensor of two BP networks, and the vacancy is filled with 1. The mapping output of two reliability functions M1 and M2 are obtained. The output results are shown in Table 10:

Step 1: The normalization coefficient $1 - k$ is calculated:

$$1 - K = 1 - \sum_{B \cap C = \varnothing}^{n} M_1(B) M_2(C)$$
$$= 1 - [M_1(\{politics\} M_2(\{Economics\}) + M_1(\{politics\}) M_2\{Military\})$$
$$+ M_1(\{Economics\}) M_2(\{Military\})]$$

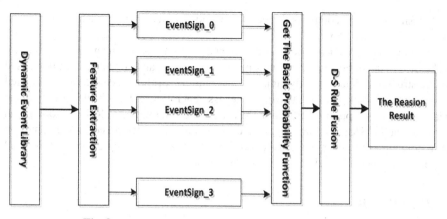

Fig. 8. BP network and DS evidence intelligence reasoning

Table 10. Mapping result of reliability function

Symptom set	M1	M2	M12
{Politics}	0.37	0.29	0.52
{Economics}	0.51	0.15	0.43
{Military}	0.12	0.57	0.48
{Θ}	0.22	0.16	0.08

Step 2: To calculate the combined reliability function of Politics: $M_{12}(\{Politics\}) = M_1 \oplus M_2(\{Politics\})$

$$= \frac{1}{1-K} \sum_{B \cap C = \{politics\}} M_1(B)M_2(C)$$
$$= \frac{1}{1-K}[M_1(\{politics\}M_2(\{politics\}) + M_1(\{politics\})M_2\{\Theta\}$$
$$+ M_1(\{\Theta\})M_2\{politics\})$$

Similarly, the values of $M_1 \oplus M_2$ ({Economics}), $M_1 \oplus M_2$ ({Military}), $M_1 \oplus M_2$ ({Θ}) can be calculated. The results are shown in Table 10. It can be seen that the four decision rules based on DS rule fusion, and the result of the reliability function shows that the main factor of inducing armed conflict between countries A and B based on the above hypothesis is political factor, which is more consistent with the actual relationship. This factor involves structured data {label_1}, and determines the keywords and spatiotemporal data in the structured data as the demand of "continuous attention" Input the data into the database, and independently coordinate the observation platform or intelligence reasoning and calculation platform of space-based or ground-based network to track and monitor the space-time and network "intelligence" targets involved in the event by the intelligent control system in the way of demand triggering. The feedback information serves as the starting point for the subsequent solution of "coordinating the relationship between countries A and B to avoid conflicts" and the iteration of event continuous tracking Point.

Through the above hypothesis and data analysis demonstration and verification, we elaborated in detail how to carry out "situation" reasoning and demand generation based on the structured data of the dynamic big data mining of the Space & Earth integrated information network, and how to trigger the task generation and feedback iteration of the heterogeneous computing platform in the Space & Earth integrated IOT network. It can be seen that the Space & Earth integrated information network based on the definition of intelligent task management and control can form a closed loop of big data demand reasoning analysis network task big data operation.

8   End Words

Based on the space-based information network structure proposed by [8, 9], this paper integrates the ground IOT information network into it, builds the integrated information network system of Space & Earth, takes the ground management and control center and the space management and control center as the centralized management and control center, and proposes an intelligent demand decision scheme of the front-end of the integrated information network task management and control based on big data and artificial intelligence technology In addition, the CMP neural network model and the fusion decision-making model proposed in this paper can be applied In the research of artificial intelligence in emotion analysis, pattern recognition and other related fields.

In this paper, the architecture, task management and control mode, and independent decision-making technology for task requirements of the Space & Earth integrated information network are discussed, but the task planning and resource planning technology for node task management and control of the satellite earth integrated network are not discussed. The next step will focus on this part to carry out docking research in the collaborative application layer of centralized management and control in the Space & Earth integrated information network space.

References

1. Na, L.Y., et al.: Analysis of spatial big data distributed storage strategy based on MySQL database. Digit. Technol. Appl. (2), 71–75 (2018)
2. Long, L.H., Liang, Y.H., Zhu, F.J., et al.: Considerations on accelerating the construction of reconnaissance and intelligence capability based on network information system. J. China Acad. Electron. Sci. 14(4), 23–25 (2019)
3. Xu, K., et al.: Application of probability theory and mathematical statistics in information theory. Sci. Technol. Inf. (10) (2008)
4. Dong, Y., Jinyu, L., et al.: Recent progresses: deep learning based acoustic models. IEEE J. Autom. Sin. 3, 18–22 (2017)
5. Chao, W.Y., et al.: Study on evaluation method of D-S evidence synthesis rules. Inf. Technol. (4) (2011)
6. Yan, D.C., et al.: Research on the application of feedforward neural network based on back propagation chaos particle swarm optimization training. Beijing University of Chemical Technology, Beijing (2013)

7. Zhao, S., Dong, X., et al.: Speech recognition based on improved LSTM deep neural network. J. Zhengzhou Univ. **5**, 31–34 (2018)
8. Peng, Q., Jiao, L.H., Zhou, L., Bin, Z., et al.: Thinking on the construction of the comprehensive management and control system of space-based information network. J. China Acad. Electron. Sci. **12**(5), 13–18 (2017)
9. Wu, M., Wu, W., Zhou, B., Lu, Z., Zhang, P., et al.: General framework of integrated information network of Space & Earth. In : Proceedings of the 12th Annual Meeting of Satellite Communication. Annual Meeting of Satellite Communication, Beijing (2016)
10. Li, G., Liu, X., Hua, G.G., et al.: Video behavior recognition algorithm based on convolutional dense network. Chin. Sci. Technol. Pap. **13**(14), 57–61 (2018)
11. Hong, W.G., Feng, C.J., et al.: Spacecraft fault diagnosis method based on BP neural network and DS evidence theory. Comput. Eng. **30**(6) (2009)

Optimization of Satellite-Ground Coverage for Space-Ground Integrated Networks Based on Discrete Global Grids

Zhu Tang[(✉)], Sudan Li[(✉)], Wenping Deng, Yongzhi Wang, and Wanrong Yu

College of Computer, National University of Defense Technology, Changsha 410073, China
tangzhu@nudt.edu.cn, nudtlsd@163.com

Abstract. Discrete global grid divides the earth's surface into approximately equal units, which is mainly used for efficient processing and visualization of the earth data. In this paper, the discrete global grid technology is introduced into the optimization of satellite-ground coverage in space-ground integrated networks. The greedy time sequence selection algorithm for satellite grid coverage is proposed to decrease the number of coverage handoff between grids and satellites, finally promotes the performance of data transmission in space-ground integrated networks.

Keywords: Discrete global grid · Space-ground integrated network · Satellite-ground coverage · Snapshot routing

1 Introduction

1.1 A Subsection Sample

At present, a variety of constellation plans with global coverage have been proposed, aiming to make the Internet more convenient for users all over the world, such as iridium and O3b constellation that are already in orbit, and Starlink constellation plan of SpaceX, oneweb constellation plan, Kuiper constellation plan of Amazon, Hongyun project and Hongyan constellation plan of China, etc. Among them, Starlink plans to launch about 12000 LEO satellites in three phases to achieve global redundant coverage, and use laser inter satellite links to build satellite networks. The blueprint created by Starlink will greatly promote the development process of the space-ground integrated network.

Because the data communication among satellites, near space vehicles, air vehicles and ground terminals are all considered in the space-ground integrated networks, the mobile and data transmission characteristics of space segment and ground segment should be considered in the initial stage of routing design. In the satellite network system with geographic location routing algorithm, the ground terminal is usually equipped with GPS module to realize fast addressing and positioning. At the same time, the system divides the earth surface into multiple geographic units by means of equal interval division of longitude and latitude. When the ground control center finds the geographic unit through the longitude and latitude information of the ground terminal, it guides

© Springer Nature Singapore Pte Ltd. 2020
Q. Yu (Ed.): SINC 2019, CCIS 1169, pp. 132–144, 2020.
https://doi.org/10.1007/978-981-15-3442-3_11

the satellite network to send the data to the satellite which is currently covering the geographical unit [10]. However, the traditional method of longitude and latitude division results in nonuniform geographical units. For example, the unit area near the equator is far larger than the unit area near the polar region, and large area units need more satellites or spot beams to cover, which increases the complexity of satellite coverage handover.

Compared with the regular icosahedron model, the discrete global grids coordinate calculation based on the regular octahedron model is simpler. The earth surface can be divided into triangular elements with approximately equal area, which are mainly used for efficient earth data processing and visualization. In this paper, the octahedron discrete global grids technology is introduced into the design of the space-ground integrated routing algorithm. Based on the predictability of satellite movement, the satellite-ground coverage time series of each grid is pre-calculated. At the same time, the snapshot routing algorithm for satellite network space segment is integrated to realize the real-time routing addressing function of the space-ground integrated network, to reduce complexity of calculating the satellite earth coverage by the traditional way. In this paper, STK and VC++ 6.0 software are used to realize the level-three subdivision of octahedral for satellite-grid coverage time series in iridium constellation, and relevant numerical analysis is given.

2 Related Works

2.1 Discrete Global Grids

In order to effectively store, extract and analyze the constantly updated global massive data, and fundamentally solve the limitations of the traditional data model, scholars put forward the spherical discrete grid model, which has the characteristics of uniform spatial location distribution, supporting multi-scale transformation, and is very suitable for the route addressing of the space-ground integrated networks. According to the method of grid generation and the shape of grid elements, the discrete global grids model can be divided into three parts: the longitude latitude spherical discrete grid model (short for latlon), the adaptive spherical discrete grid model and the polyhedron spherical discrete grid model [1].

The longitude latitude spherical discrete grid model is the earliest and used most widely. The typical models include VGIS (Virtual GIS) [2] of Georgia Institute of technology, VPE (Virtual Planetary Exploration) [3] of NASA, and the SIMG (Spatial Information Mulit-Grid) technology adapted to grid computing environment of Li Deren in China [4, 5]. The longitude latitude spherical discrete grid model meets people's use habits and can be simple transformed to other coordinate systems, but there are also some shortcomings, such as the area of grid between high and low latitude is inconsistent, the coordinates of North and south poles are changing and oscillating, the data density of high and low latitude is inconsistent with the coverage area, and it is difficult to establish regional spatial index.

Based on the solid elements on the sphere, the adaptive spherical discrete grid model divides the spherical elements according to some characteristics of the solid elements. Its main theory is the Voronoi diagram with dynamic stability characteristics, and the main schemes are the LOD model of the global digital elevation model data [6], the spatial data management tool vordll [7] of Laval University in Canada, etc. Due to the special shape of the adaptive grid, it has more flexibility than the regular grid, but it is also difficult to achieve recursive segmentation and multi-scale massive data association.

The spherical discrete grid model of regular polyhedron mainly projects the shapes of regular tetrahedron (4 equilateral triangles), regular hexahedron (6 squares), regular octahedron (8 equilateral triangles), regular dodecahedron (12 pentagons) and regular icosahedron (20 equilateral triangles) on the sphere to produce spherical polygons with the same shape. It overcomes the defects of the non-uniformity and the singularity of the two poles of the longitude latitude grid model, and has the characteristics of good stability and approximately uniform division in the global scope. At the same time, it supports the direct positioning of the grid on the global surface and the mutual transformation of the sphere and the geographical coordinates. At present, the octahedron QTM (Quad triangle mesh) [8] and the icosahedron sphere [9] discrete grid models have become effective tools for building a global hierarchical grid model.

2.2 Spatial Network Geographic Information Routing

Geographic information routing algorithm has a long history in the space-ground integrated networks. This kind of algorithm usually divides the ground surface into regions of the same size, and assigns a fixed logical address to each region. Each packet carries the logical address as the source/target address. The satellite node judges the user's region according to the logical address carried by the packet, and then forwards the packet to the satellite covering the user's region using the route based on geographical location, and finally forwards it to the user. The actual area division can be different according to the different application scenarios, and the hierarchical structure can also be adopted for area division. Typical satellite network geographic routing algorithms include DGRA algorithm [10], SIPR algorithm [11], etc. However, the scheme proposed mainly uses the way of binding the ground terminal to the access satellite to realize the location management of the terminal. All the binding information will be continuously updated in the ground control center with the satellite moving, resulting in a large amount of data update costs.

To solve this problem, Tsunoda [12] proposed a handover independent mobile management mechanism for IP/LEO satellite network. This mechanism can shield the impact of satellite handover on terminal addressing by storing the geographic location information of the ground terminal, effectively reducing the update cost of binding information. Meanwhile they take the orbit information of satellite into account to solve the confusion problem of the last hop selection of adjacent satellites [13], improving the accuracy of ground terminal addressing. However, although GPS positioning information is used to record the location of the ground terminal, and the ground is divided into multiple cells to determine the coverage area of the satellite, the real-time coverage sequence of the satellite network to the ground unit is not given, so it is difficult to quickly locate the satellite covered by the ground unit. In recent years, although software defined network [17], network function virtualization [18], software defined radio [19] and other new network technologies have been widely introduced into satellite network, but in general, they have not changed the essential law of satellite ground coverage, and optimizing the performance of satellite ground coverage is still very important for improving the networking performance of the above-mentioned new technologies in the space-ground integrated network.

Therefore, this paper intends to introduce the discrete global grids technology into the space-ground integrated networks, and realize the reasonable satellite ground coverage distribution through the uniform and hierarchical spherical division. At the same time, based on the regularity and predictability of the satellite movement, the coverage time series of the satellite to the ground grid can be calculated in advance, to improve the fast addressing ability independent of the satellite-ground handover.

3 Introduction to Discrete Global Grid

3.1 Division of Discrete Global Grid

In order to achieve the uniform coverage for the ground, this paper proposes to use the regular octahedron discrete grid model to divide the earth surface. The vertex of the octahedron coincides with the main position points (including poles) of the earth. Take the original spherical triangle of the octahedron ABC for example, vertex A coincides with the pole, and the other two vertices are located at the equator (Fig. 1).

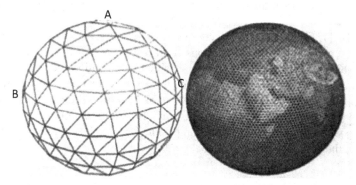

Fig. 1. Schematic diagram of level 3 (left) and level 6 (right) division of octahedron [1]

Literature [1] compared the basic properties of discrete global grids based on octahedron at different levels of subdivision (n = 0–13). It was found that with the increase of subdivision levels, the ratio (the ratio of the maximum to the minimum edge of the spherical triangle with the largest deformation in each level, the ratio of the maximum to the minimum angle, the ratio of the maximum to the minimum perimeter as well as the ratio of the maximum to the minimum sphere triangle area) increases gradually from 1, which represents the uniform division, and then tends to be stable. At last, the ratios keep 1.559394, 2, 1.370207 and 2.105921 respectively. Furthermore, the closer to the vertex of the octahedron, the larger the deformation is, and the farther away from the vertex, the smaller the deformation is. Although the multi-level discrete grid based on the regular octahedron is not a uniform sphere division, it is still superior to the longitude latitude sphere discrete grid model and the adaptive sphere discrete grid model. Therefore, the

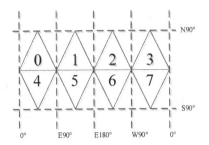

Fig. 2. Coding method of a0

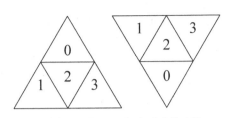

Fig. 3. Coding method of ai ($i \neq 0$)

spherical discrete grid model based on the regular octahedron division is employed in this paper. At the same time, considering the orbit altitude of the experiment satellite constellation, the spherical area is divided by the level $n = 2$.

3.2 Coding of Discrete Global Grid

In order to accurately represent each discrete grid, a unique code is assigned to each grid. With this code, not only the location of the grid, but also the subdivision level (or resolution) of the grid are reflected, ensuring that any area on the ground can be mapped into a given grid.

Referred to [1, 14], QTM grid can code any grid as follows, that is, $q = a0a1a2a3\ldots an$, where a0 is represented by eight fractions from 0 to 7, and ai ($i = 1, 2, 3, \ldots$) are represented by four fractions of 0–3. The specific determination method is shown in Figs. 2 and 3. In order to save storage space, binary code can be used for storage in practical application, that is, 3-bit binary code (000-111) is used for a0 level, and 2-bit binary code (00-11) is used for subsequent level. Therefore, the QTM code used is a binary string with uncertain length. For example, the QTM code of level 2 grid 'a0a0a1' is mapped to the binary code '000 00 01'.

3.3 Coordinate Calculation of Discrete Global Grid

In order to facilitate the use and intuitive analysis, in most cases, we need to convert the coding of octahedral discrete global grids into longitude and latitude coordinates for positioning and operation. The existing conversion methods mainly include ETP projection method, zot projection method and row/column approximation method. Among them, row/column approximation method [15] recursively approximates the coding according to the row and column of QTM in a certain direction. This method has fast operation speed and good accuracy related to the size of the grid, but has global stability characteristics. Therefore, this paper uses this method to calculate the longitude and latitude coordinates of the discrete grid.

In order to simplify the calculation complexity of satellite ground coverage, this paper intends to perform satellite ground coverage strategy calculation based on the triangular vertex of each grid. Therefore, after completing the discrete global grids division, we need to calculate the longitude and latitude coordinates of each grid vertex. This paper mainly uses the method described in reference [1] to convert QTM code to longitude and latitude, and adds the definition and calculation method of grid vertex. This method transfers by means of triangle unit coordinate system, that is, first convert QTM code to triangle unit coordinate, and then convert from triangle unit coordinate to longitude and latitude. In reverse, we firstly convert longitude and latitude to triangle unit coordinate, and then, the triangle unit coordinate is converted to QTM code.

The coordinate system of the triangle unit vertex is shown in Fig. 4. The data pairs in each unit represent the coordinates of triangle unit. The origin of the coordinate system is $(0, 0)$ unit, the Y axis lies with the leftmost column unit, and the X axis extends along the unit with the same y value. The coordinate of each unit is represented as (y, x). The larger the value of y is, the larger the value range of x is. For example, when $y = 1$, x can be taken as $[0, 1, 2]$. Any coordinate (y, x) of level n partition satisfies the relation $\{(y, x) \mid y \in [0, 2n - 1], x \in [0, 2y], n > 0\}$.

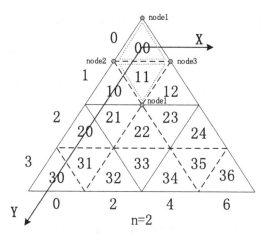

Fig. 4. Coordinate system of the triangular unit vertex

(1) Transformation from triangle unit coordinates to QTM code

The process of transforming triangle unit coordinates (y, x) into QTM code is to make $m = 1 (1 \leq m \leq n)$, and calculate the QTM code layer by layer from level 1 depth to level n depth. The algorithm is as follows [1]:

   (1) Judge the relative position of the four parts of the parent unit for the current unit. If $y < 2^{(n-m)}$, it will be in the 0th sub unit, and the corresponding code of the depth is $am = 0$ and go to (3);
   (2) If $x \leq 2y - 2^{(n-m+1)}$, code $am = 1$ and $y = y - 2^{(n-m)}$, go to (3); if $x < 2^{(n-m+1)}$, in 2nd sub unit, code $am = 2$ and $x = x - 2y + 2^{(n-m+1)} - 1$, $y = 2^{(n-m+1)} - y - 1$, go to (3); otherwise, in 3rd sub unit, code $am = 3$, $x = x - 2(n - m + 1)$ and $y = y - 2(n - m)$, turn to 3;
   (3) If $m > n$, the conversion process is terminated and exited; otherwise, $m = m + 1$, and go to (1) for the next depth calculation.

(2) Transformation from triangle unit coordinate to longitude latitude coordinate

Suppose that the up vertex of grid cell a0 of level 0 is at the north pole (90° north latitude), the longitude and latitude of the left vertex are (0°, 0°), the right vertex is (90°, 0°), and the sub unit A is obtained by level n subdivision. Suppose that the triangle unit coordinates of A are (y, x), and the maximum Y-axis coordinates of level n subdivision are $2^n - 1$ (starting from 0), so the latitude occupied by each unit is $90/2^n$. Since there are 2y sub units in the row A lies in, the longitude width took up by each triangle unit is 90/2y.

The vertex coordinate calculation formula of level 0 grid unit a0 is shown in Table 1, where the grid a0a0a0 triangle faces up, and the up vertex is node1, lower left vertex is node2, the lower right vertex is node3, while the grid a0a1a1 triangle faces down, and the up vertex is node1, the upper left point is node2, and the upper right point is node3.

Table 1. Example of grid vertex coordinate calculation in a0 (level 2 division)

No	DGG Code	DGG Vertex	Lat Calc	Long Calc
1	a0a0a0 (triangle pointing to poles)	node1	$90.0 * (1 - y/2^n)$	$90.0 * x/(2 * y)$
2		node2	$90.0 * (1 - (y + 1)/2^n)$	$90.0 * x/(2*y + 2)$
3		node3	$90.0 * (1 - (y + 1)/2^n)$	$90.0 * (x + 2)/(2 * y + 2)$
4	a0a1a1 (triangle pointing to the equator)	node1	$90.0 * (1 - (y + 1)/2^n)$	$90.0 * (x + 1)/(2 * y + 2)$
5		node2	$90.0 * (1 - y/2^n)$	$90.0 * (x - 1)/(2 * y)$
6		node3	$90.0 * (1 - y/2^n)$	$90.0 * (x + 1)/(2 * y)$

4 Satellite Ground Coverage Division Optimization and Addressing Procedure

4.1 Greedy Selection Algorithm for Grid-Satellite Coverage Time Series

In order to optimize the performance of satellite-ground coverage division, this paper proposes the greedy selection algorithm for satellite-grid coverage time series in Table 2. The algorithm sets the total coverage time as D, the coverage time set C_i of grid i as 0, and each time selects the satellite with the longest remaining coverage time to all the vertexes of grid i. After selecting this coverage, the coverage time C_i of grid i is extended, and the switching time is set according to the sequence of the adjacent coverage time series. For example, the first selected satellite needs to last until it cannot be covered and then switch to the next selected satellite to reduce the number of handovers. When the coverage time C_i of grid i reaches the total coverage time D or all the coverage has been selected, the selection operation is completed and the algorithm ends.

Table 2. Greedy algorithm for satellite-grid coverage time series selection

Input: the coverage time series set Mi of grid i for each satellite, the target coverage duration D of grid, and the coverage duration calculation function C based on the coverage time series between grid and satellite;
Output: satellite coverage time series A_i of grid i.

Initialization: for grid i, total coverage time $C_i = 0$
1   while $C_i \,!= D$
2   for each m_{ij} in M_i
3            $A_i' = A_i \cup \{m_{ij}\}$;
4   if $C(A_i') > C_{temp}$
5                $k = j$
6                $C_{temp} = C(A_i')$
7   endfor
8        $A_i = A_i \cup \{m_{ik}\}$
9        $M_i = M_i - \{m_{ik}\}$
10   endwhile
11   return A_i

4.2 End to End Addressing Process Based on Discrete Global Grid

Before sending data, the source and destination terminals need to register with the currently covered satellite. Assumed that each terminal is equipped with a global positioning module and can report its own longitude and latitude position attribute, the satellite network can bind the terminal to the corresponding grid unit code through coordinate system conversion based on the longitude and latitude information. When the access satellite of the source ground terminal obtains the grid code of the destination ground terminal through the ground control center, the access satellite needs to address based on the internal route of the satellite network segment and the satellite-ground coverage route. At the same time, based on the periodicity and predictability of the satellite movement, the end-to-end transmission path including the satellite segment and the satellite ground segment are integrated into consideration, to improve the routing performance and availability.

For example, when the source terminal A needs to send data to the destination terminal B, the source terminal A first reports the terminal identification of the destination terminal to the ground network operation and control center (NOCC) through the current coverage satellite S_A, and obtains the corresponding grid unit code Q_B of the destination terminal by querying the terminal registration database. The data of source terminal A can be real-time addressed and forwarded in satellite network based on the grid code. The route of space segment in satellite network can adopt the snapshot route described in reference [16]. At this time, the route of space segment should consider the snapshot route table switching caused by the inter satellite link connection and the destination satellite change caused by the grid coverage switching, both of which can be based on the precise time synchronization function for the whole network. When the data reaches satellite S_B which is covering the grid of the destination terminal, satellite S_B selects corresponding transponder and downlink based on the grid code to transmit the data to the destination terminal B.

5 Performance Analysis

5.1 Numerical Analysis of Grid Coverage by Satellites

In this paper, iridium constellation and level 3 normal octahedron discrete global grids model are used for testing. There are no restrictions on the elevation angle and azimuth angle of ground terminal, and no restrictions on the elevation and azimuth angle of satellite downlink coverage. The number of visibility coverage time series comparison between the original one and the greedy algorithm is shown in Fig. 5, where the test time is an orbit period (about 6027 s). In this figure, Y-axis represents the number of satellite coverage time series of each grid in the test time, and X-axis represents the sequence number corresponding to the grid. The triangle code of the grid is defined as (y, x), the sequence number is calculated as follows:

$$SeqNo = \sum_{i=1}^{y-1} (2i - 1) + x + 1$$

It can be seen from Fig. 5 that the average number of original visibility coverage in an orbital period of level 3 subdivision (DGG, which is 23.73) is lower than the longitude latitude grid division model with 10° (latlon10, 34.13), and the longitude latitude grid division model with 12° (latlon12, 31.48). Although the number of greedy selected coverage of DGG (12.19) is larger than latlon10 (11.77) model and smaller than latlon12 (12.36) model, some grid of the latlon12 model near the equator could not be continuously covered by the Iridium system, which will badly affect the satellite-ground communication.

At the same time, we can also see that the closer the polar area is, the more frequent the switching between the grid and the satellite, resulting in a very large number of coverage time series. For example, the average original coverage number in one orbit cycle of grid code a0a0a0a3, a1a0a0a2, a2a0a0a1, a3a0a0a3, a4a0a0a3, a5a0a0a2, a6a0a0a2, a7a0a0a1 (serial number as 4, 67, 130, 196, 260, 323, 387, 450) is about 71.5, which is much higher than that of greedy algorithm.

Finally, the standard deviation of coverage number of DGG model (13.92 and 1.22) is smaller than latlon10 model (21.40 and 1.27) and latlon12 model (19.99 and 1.68). The stabilization of satellite-grid coverage of DGG is better than longitude latitude division model.

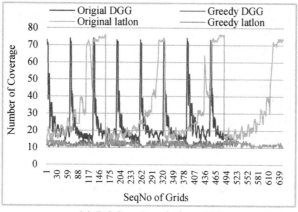

(a) DGG n=3 vs latlon=10°

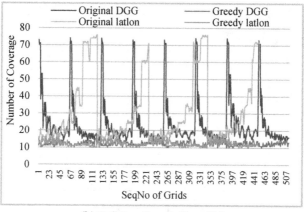

(b) DGG n=3 vs latlon=12°

Fig. 5. Comparison of the number of coverages between discrete global grids and longitude latitude division model (6027 s)

5.2 Analysis of Satellite-Grid Coverage Time

The duration of satellite coverage sequence of some grid in a single orbital period (6027 s) is shown in Fig. 6. It can be seen from the figure that the satellite coverage time of the grid has changed in stages, which is related to the changing relative position between the satellite and the earth, and the decrease of the coverage time of the last sequence is due to the end of the test time. At the same time, we can see that DGG model owns much closer curves than latlon model, which is because the average coverage time of DGG model is more stable than latlon model.

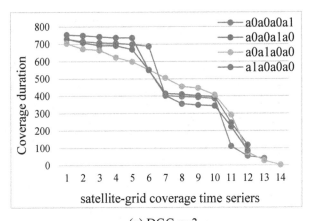

(a) DGG n=3

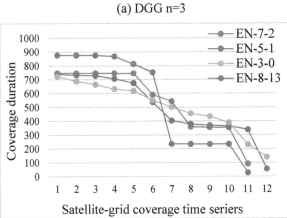

(b) latlon = 10°

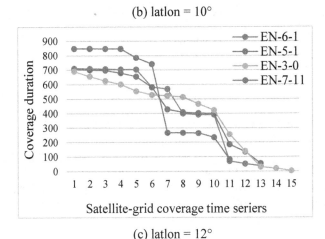

(c) latlon = 12°

Fig. 6. Analysis of satellite coverage time of some units in the case of discrete global grids and division of longitude and latitude (6027 s)

6 Concluding Remarks

Aiming at the complex relative motion between the space segment and the ground segment of the satellite-ground integrated network, this paper introduces the discrete global grids technology into the design of the satellite-ground integrated routing algorithm. This method divides the earth surface into triangular grid units of approximate size by level n subdivision, and pre-plan the satellite coverage time series of each grid in advance based on the predictability of satellite movement, and propose the greedy algorithm to reduce the handover numbers between satellites and grids. In this paper, STK software is used to simulate the coverage scene of iridium constellation to the level 3 normal octahedron discrete global grids, and the effectiveness of the algorithm for the grid satellite coverage time series generation and greedy selection is verified.

References

1. Zhang, S.: Researeh of Global Remote Sensing Image Browser System Based on Octahedron Discrete Global Grids Model, East China Normal University (2009)
2. Faust, N., Ribarsky, W., Jiang, T., et al.: Real-time global data model for the digital earth. In: Proceedings of the International Conference on Discrete Global Grids (2000)
3. Hitchner, L.: Virtual planetary exploration: a very large virtual environment. In: ACM SIGGRAPH 1992, Tutorial on Implementing Immersive Virtual Environments (1992)
4. Li, D., Cui, W.: Geographic ontology and SIMG. Acta Geodaetica Cartogr. Sin. **35**(02), 143–148 (2006)
5. Li, D., Shao, Z., Zhu, X.: Spatial information multi-grid and its typical application. Geomat. Inf. Sci. Wuhan Univ. **29**(11), 945–950 (2004)
6. Kolar, J., Dgi, C.: Representation of geographic terrain surface using global indexing. In: International Conference on Geoinformatics Geospatial Information Research: Bridging the Pacific & Atlantic (2008)
7. Gold, C., Mostafavi, M.: Towards the global GIS. ISPRS J. Photogramm. Remote. Sens. **55**(3), 150–163 (2000)
8. Dutton, G.: Encoding and handling geospatial data with hierarchical triangular meshes. In: Symposium on Spatial Data Handling (1996)
9. Gong, J., Tong, X., Zhang, Y., et al.: Research on generating algorithm and software model of discrete global grid systems. Acta Geodaetica Cartogr. Sin. **36**(02), 187–191 (2007)
10. Henderson, T., Katz, R.: On distributed, geographic-based packet routing for LEO satellite networks. In: Proceedings of the Global Telecommunications Conference 2000 (GLOBECOM 2000), pp. 1119–1123 (2000)
11. Hashimoto, Y., Sarikaya, B.: Design of IP-based routing in a LEO satellite network. In: The 3rd International Workshop on Satellite-Based Information Services, Mobicom 1998, pp. 1–6 (1998)
12. Tsunoda, H., Ohta, K., Kato, N., et al.: Supporting IP/LEO satellite networks by handover-independent IP mobility management. IEEE J. Sel. Areas Commun. **22**(2), 300–307 (2004)
13. Hiroshi, T., Kohei, O., Nei, K., Yoshiaki, N.: Geographical and orbital information based mobility management to overcome last-hop ambiguity over IP/LEO satellite networks. In: IEEE International Conference on Communications (2006)
14. Zhang, S., Wu, J., Gan, J.: Research on triangle subdivision and cell search based on equilateral octahedron. In: Geoinformatics & Joint Conference on GIS & Built Environment: Advanced Spatial Data Models & Analyses. International Society for Optics and Photonics (2008)

15. Zhao, X., Chen, J.: Fast translating algorithm between QTM code and longitude/latitude coordination. Acta Geodaetica Cartogr. Sin. **32**(3), 272–277 (2003)
16. Zhu, T., Wanrong, Y., Zhenqian, F., Wei, H., Baokang, Z., Chunqing, W.: Rollback traffic avoidance for snapshot routing algorithm in cyclic mobile networks. In: The 10th IEEE International Conference on Networking, Architecture, and Storage (NAS 2015), pp. 151–157 (2015)
17. Bi, Y., Han, G., Xu, S., et al.: Software defined space-terrestrial integrated networks: architecture, challenges, and solutions. IEEE Network **33**(1), 22–28 (2019)
18. Li, T., Zhou, H., Luo, H., Yu, S.: Service: a software defined framework for integrated space-terrestrial satellite communication. IEEE Trans. Mob. Comput. **PP**(99), 1 (2017)
19. Frank P., Fatemeh A., Radhika R., William E.: Software defined radio implementation of DS-CDMA in inter-satellite communications for small satellites. In: 2015 IEEE International Conference on Wireless for Space and Extreme Environments (WiSEE), pp. 1–6 (2015)

Research on Information Network Invulnerability of Space-Based Early Warning System Based on Data Transmission

Lifang Liu[1], Yan Wang[1(✉)], Wei Xiong[2], Jialin Hou[1], and Xiaogang Qi[3]

[1] School of Computer Science and Technology, Xidian University, Xi'an 710071, Shaanxi, China
1751611337@qq.com
[2] Science and Technology on Complex Electronic System Simulation Laboratory, Space Engineering University, Beijing 101416, China
[3] School of Mathematics and Statistics, Xidian University, Xi'an 710126, Shaanxi, China

Abstract. In order to ensure the reliability of data transmission and meet the QoS (quality of service) requirements of different users, this paper analyzes the invulnerability of space-based early-warning system information network based on data transmission, taking the space-based early-warning system information network as the carrier. First of all, this paper classifies the network invulnerability. Secondly, according to the damage of the network under different circumstances, the author designs the corresponding routing strategies of invulnerability and compares it with the traditional routing algorithm. The simulation results show that the routing strategies of invulnerability proposed in this paper not only satisfies QoS requirements of different users, but also has good performance in packet loss rate, end-to-end delay and other aspects because it considers many aspects of network damage.

Keywords: Invulnerability · QoS · Space-based early warning system

1 Introduction

The space-based early warning system is the main component of the strategic early warning system. It collects and transmits data through the space-earth integrated network [1, 2]. The main function of the space-based early warning system is taking advantage of space detection to discover, identify, track and monitor enemy ballistic missiles. By measuring the relevant parameters of incoming missiles, the information such as missile landing point, launch point, incoming time, flight trajectory, threat degree and interceptability can be inferred to provide various support information needed for interception and counterattack. The space-based early warning system usually uses high and low orbit satellite network, which is generally composed of space satellite early warning

Project supported by the National Natural Science Foundation of China (Grants No. 61877067), Equipment sector fund (Grants No. 61420100201162010002-2).

© Springer Nature Singapore Pte Ltd. 2020
Q. Yu (Ed.): SINC 2019, CCIS 1169, pp. 145–162, 2020.
https://doi.org/10.1007/978-981-15-3442-3_12

network, space information transmission network and ground network. This system, which is not affected by ground curvature, covers a large scale of ground and space, and only a limited number of satellites are needed to achieve global coverage. In the space information transmission network of space-based early warning system, a harsh and strong electromagnetic environment in which space satellite nodes locate results in the failure of satellite nodes and links, and changes the network topology. Furthermore, the number of data packets cannot be transmitted correctly through any end-to-end paths of the nodes or links. This increases data packet loss rate and makes network more vulnerable to human attacks. Eventually, a large number of malicious packets flow in network instantaneously, leading to network congestion or even paralysis which makes it unable to provide normal service. Therefore, it is very important to study the invulnerability of networks in the context of space-based early warning systems.

Nowadays, the research on the invulnerability of space-based early warning system network can be extended to three main questions: (1) how to measure the invulnerability of networks; (2) how to design a network with invulnerability; (3) how to optimize the capacity of the network. Among them, the measurement of network invulnerability refers to how to quantitatively analyze the level of network invulnerability. Currently, it is mainly based on two major theories: graph theory and statistical physics [3]. The network invulnerability evaluation based on graph theory includes connectivity [4], tenacity [5], integrity, dispersion and network efficiency. Albert [6] proposed a method to measure the maximum connected sub-graph and the average path length. This method focuses on the impact of the topology on the invulnerability through random or deliberate attacks on the network topology. Huang [7] proposed a network invulnerability measure based on natural connectivity on the basis of the common invulnerability measures. Statistical physics-based network invulnerability assessment studies mostly use random failure and intentional attack failure strategies to study network invulnerability. The design of network invulnerability refers to how to design a network that meets a certain degree of invulnerability. The main means is to equip properly the network with enough backup devices, but not unrestrainedly. Reference [8] aiming at the blockage of some nodes in the transmission process of multi-layer satellite network combined with MEO and LEO, proposes a multi-coverage model of upper satellite to lower satellite network to increase the number of links between different network layers and realize the balanced distribution of transmission traffic. In Ref. [9], the Teledesic constellation system is adopted to introduce the idea of multi-path redundant transmission in multi-path parallel transmission, and solves the problem of low reliability of multi-path parallel transmission and meets the transmission demand of high-bandwidth data. The optimization of network capacity refers to how to satisfy the logic invulnerability under the premise that the network topology is invulnerable. The main research is how to arrange reasonable communication paths and backup paths for existing communication services in the network, and allocate bandwidth rationally on the corresponding paths, so as to reduce the construction cost as much as possible while network satisfies the transmission capacity and invulnerability required. Reference [10] sets the time threshold of transmission. According to whether the time of data transmission exceeds the time threshold, different layers of networks are selected as transmission paths by exploiting the strengths of different layers of satellites. Reference [11] proposes a traffic adaptive algorithm based on transmission distance, which determines at which satellite layer the data is transmitted by judging the transmission distance. Reference [12] proposes an intelligent routing strategy based on traffic lights, and a group of traffic lights is used to indicate the congestion status of the current node and the next node. When the group reaches its destination along the pre-calculated route, the route can be dynamically adjusted according to the real-time color of the traffic lights of each intermediate node, Finally, the approximate best transmission path can be obtained.

Aiming at the information network invulnerability of space-based early warning system, this paper analyzes the data transmission of network under different circumstances, such as port congestion, node congestion, link failure and node failure, and designs different invulnerability strategies to optimize network capacity and improve network invulnerability. The structure of this paper is as follows: Sect. 2 introduces the space information transmission network model of space-based early warning system and classifies the network invulnerability; Sect. 3 introduces some details different network invulnerability technology under different circumstances; In Sect. 4, the network invulnerability in different situation is analyzed; Sect. 5 summarizes the whole passage.

2   System Model and Network Invulnerability Analysis

2.1   System Model

The network structure of space-based early warning system mainly consists of three parts, namely, space satellite early warning network, space information transmission network and ground network. This paper mainly studies the invulnerability of network during data transmission, so space information transmission network and ground network are mainly used. The system model is shown in Fig. 1:

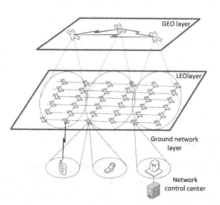

Fig. 1. Schematic diagram of system model

We use a two-layer satellite network consisting of 3 GEO satellites and 66 LEO satellites. The ground network includes some sender nodes, receiver nodes and network control centers. Three GEO satellites are connected to each other, and their main functions are to manage the network, flood the information of the failed nodes, and also divert the flow when LEO satellite is congested. Iridium satellite model is adopted for LEO satellite, which mainly carries out network communication, and there is no communication between polar regions. The ground transceiver node mainly realizes sending and receiving all kinds of data packets. The ground network control center is used to calculate and update the route and send the calculated path information to the corresponding LEO satellite.

With the movement of the satellite, the distance between the inter-orbit satellites in the satellite network model is constant, but inter-orbit links and inter-layer links changes, which will trigger route reconstruction. So the distance between adjacent satellites in inter-orbit can be calculated as follows

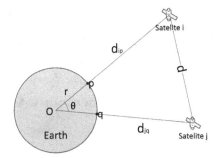

Fig. 2. Schematic diagram of inter-star distance

As shown in Fig. 2, suppose the radius of the earth is r, the angle between satellite i and satellite j is θ, p and q are the points under the satellite i and satellite j respectively. d_{jq} represents the distance between satellite j and point q, The latitude and longitude of point p and q are (lon_p, lat_p), (lon_q, lat_q). The distance d is calculated as follows:

$$\theta = \arccos(\cos(lat_p)\cos(lat_q)\cos(lon_q - lon_p) \\ + \sin(lat_p)\sin(lat_q)) \tag{1}$$

$$d = \sqrt{d_{ip}^2 + d_{jq}^2 - 2d_{ip}d_{jq}\cos\theta} \tag{2}$$

2.2 Classification of Network Invulnerability

In this paper, the network invulnerability technology is divided into two aspects: (1) network invulnerability in logic, (2) network invulnerability in physics.

Network invulnerability in logic refers to the optimization of network capacity. When there are a large number of packets in the network, it is necessary to design an effective invulnerable routing algorithm to arrange a reasonable communication path and backup path, and allocate the bandwidth reasonably, so that the original communication services can meet the QoS requirements of different users to the greatest extent. Network invulnerability in logic is divided into in the case of port congestion and node congestion.

Port congestion refers to data flow meet at the node and the orientations of are concentrated, which leads to the congestion of a certain data transmission port of the node. Node congestion indicates that the node is of high importance. In this case, to change the congestion in the network, the first way is to divide the data through other nodes in the network, and the second way is to replace the node forwarding device to improve the adaptability of the space-based early warning system.

Network invulnerability in physics refers to how to re-route the path according to the real-time network conditions when the satellite node or link is attacked by others, so that the services within the network can still be unaffected and meet the QoS requirements of different users simultaneously. Network invulnerability in physics can be divided into network invulnerability in the case of link failure and network invulnerability in the case of node failure.

3   Research on Network Invulnerability Technology

3.1   Network Adaptation Model in the Case of Congestion

The node, during the transmission progress, always chooses a different path according to the different congestion extent. In this section, we first present a method to calculate path. Then, we analyze the congestion of node and port in the network, and design an invulnerability routing scheme. Finally, we conduct a simulation to test the performance of this scheme.

(i)   Path calculation

Assume that the transmission delay from the source node to the destination node consists of three parts. (1) the delay from the source node to the access satellite; (2) the delay from the accessing satellite of the source node (s_leo) to the accessing satellite of the destination node (d_leo); (3) the delay from d_leo to the destination node. Then, the delay from the source node to the destination node can be obtained by

$$Delay_{(s,d)} = Delay_{(s,s_{leo})} + \sum_{i=1}^{n} Delay_{(i,j)} + Delay_{(d_{leo},d)} \tag{3}$$

At the beginning of the time slice, according to the adjacency between nodes, we first construct an adjacency matrix TotalMatrix to store the distance between the adjacent satellites. If a satellite locates in the polar region, we mark it as unreachable. Secondly, the Dijkstra algorithm is used to calculate the optimal path for each LEO satellite node to other nodes. Let $L_{(i,j)}$ denote the optimal path from node V_i to node V_j and $LS_{(i,j)} = \{E_{(m,n)} | E_{(m,n)} \in L_{(i,j)}\}$ denote the link set of the optimal path, where $E_{(m,n)}$ is an inter-satellite link from the optimal path. Then, we set the adjacency attribute of $LS_{(i,j)}$ in TotalMatrix as unreachable, and recalculate the optimal path from node V_i to node V_j to get the sub-optimal path $\Gamma_{(i,j)}$. Through the above steps, the optimal routing table and the sub-optimal routing table for each LEO satellite node to transmit data can be obtained.

As shown in the Fig. 3, the red line is the optimal path calculated from the current node to the destination node, and the green line is the sub-optimal path. The two paths don't overlap

current node

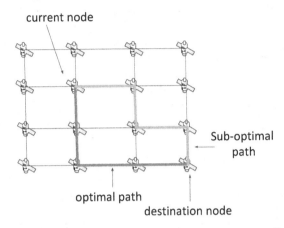

Sub-optimal path

optimal path

destination node

Fig. 3. Optimal path and sub-optimal path (Color figure online)

(ii) **Network adaptation model in the case of port congestion**

When the node is transmitting data, it usually choose the port of optimal path to connect. If congestion occurs at that port, transferring data through other ports at that node should be considered. Each port of the node, which transmits incoming packets in order, can be viewed as a queue model. However, since the queue model cannot meet the QoS guarantee of users, we need to make changes. We first classify the services, then set a threshold for each buffer queue, and finally adopt different transport modes for different types of services. This can not only ensure the QoS but also reduce packet loss.

In this paper, services are divided into three types, namely C1, C2 and C3. C1 are mainly some voice data, which are very sensitive to delay requirements. C2 are mainly pictures and video data, which have low delay requirements. C3 are mainly some file data, which have no delay requirements. In this study, Services of type C1 and C2 are transferred first, C3 are finally considered.

Suppose the node has four ports: 1, 2, 3, 4 $Q(n)(n = 1, 2, 3, 4)$ is the packet occupancy of port n. In order to better distinguish port congestion, two thresholds M_1, $M_2(M_1 < M_2)$ are set for $Q(n)$. According to the congestion situation of port, the capacity of port to handle services as follows (Table 1):

M_1, M_2 need to be given the boundary limit. For example, at a certain moment, the packet occupancy of a port is close to M_1, but it is not updated until time t, so the status of non-congestion is always displayed. Due to the delay in updating, the port received a large number of packets during this period of time, which directly caused the occupation rate of port packets to exceed b, resulting in congestion. Similarly, if the

Table 1. The ability of port to handle services

Data occupancy of port n	Congestion situation	Services that can be handled
$[0, M_1]$	No	C1, C2, C3
$[M_1, M_2]$	Slight	C1, C2
$[M_2, 1]$	Heavy	C1

packet occupancy of a port is close to M_2, the port received a large number of packets at time Δt causes the port occupancy to exceed it. Therefore, the following restrictions need to be met:

$$\Delta t \cdot (Input1 - Output1) \leq (M_2 - M_1) \cdot n \cdot s \tag{4}$$

$$\Delta t \cdot (Input2 - Output2) \leq (1 - M_2) \cdot n \cdot s \tag{5}$$

From (4) and (5), we get

$$M_1 \leq 1 - \frac{\Delta t[(Input1 - Output1) + (Input2 - Output2)]}{n \cdot s} \tag{6}$$

$$M_2 \leq 1 - \frac{\Delta t(Input2 - Output2)}{n \cdot s} \tag{7}$$

So we set

$$M_1 = 1 - \frac{\Delta t[(Input1 - Output1) + (Input2 - Output2)]}{n \cdot s} \tag{8}$$

$$M_2 = 1 - \frac{\Delta t(Input2 - Output2)}{n \cdot s} \tag{9}$$

Where Δt represents the updated time interval, $Input1$ represents the input traffic at time Δt boundary M_1, $Output1$ represents the output traffic at time Δt boundary M_1, $Input2$ represents the input traffic at time Δt boundary M_2, and $Output2$ represents the output traffic at time Δt boundary M_2, n is the number of packets and s is the size of each packet.

Figure 4 is about the data transfers. Satellite S sent packets C1, C2 and C3 to the satellite m. The satellite m including four ports (regardless of the port which is connected with GEO). Each of them has a buffer queue, and is connected with a LEO satellite. Satellite m has three choices satellite i, satellite j and satellite k when making the next hop selection. Suppose that satellite i is the next hop of the optimal path selected by DSP algorithm, and satellite j is the next hop of the sub-optimal path. When single port congestion occurs, the congestion control scheme is as follows:

Step 1: When the packet arrives at satellite m, it first checks the congestion of each port of m. If all ports are congested, the packet is discarded directly. Otherwise, do the following steps.

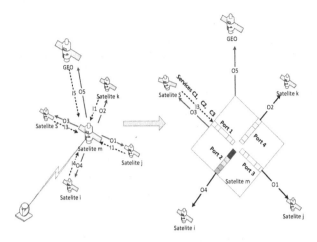

Fig. 4. Data transmission diagram of port

Step 2: Choose port 2, which connects to the shortest path for transmission. If the packet occupancy of this port is $[0, M_1]$, it means that there is no congestion on this port. At this time, all three types of packets are transmitted from this port.

Step 3: If the packet occupancy of port 2 is $[M_1, M_2]$, it means that the port is slightly congested. So just C1 and C2 packets are transferred from this port, and C3 packets are transferred from other ports.

Step 4: If the packet occupancy of port 2 is $[M_2, 1]$, it means that the port is heavily congested. So only C1 packets are transmitted from this port, and C2 packets are transmitted from port 3 which connected with the sub-optimal path, and C3 packets are transmitted from the remaining port.

(iii) **Network adaptation model in the case of node congestion**

Querying congestion on each port can effectively reduce packet loss rate and divert different services from different ports. This section describes the measures that solve how to handle packets when all ports are congested.

Each node has buffer queues in four transmission directions. The share of each buffer queue is denoted by $Q(n)$, and the share of all buffer queues of the whole node is denoted by $AQ(n)$. Similarly, we set two thresholds AM_1, AM_2 for $AQ(n)$, and according to the congestion situation of nodes, the capacity of nodes to handle business is as follows (Table 2):

The satellite's state changes, it notifies its neighbors immediately. This allows the satellite to see the status of its neighbors in real time. Figure 5 is the schematic diagram of data transmission. Satellite s sends packets C1, C2 and C3 to satellite m which connects with satellite i, satellite j, satellite k and GEO satellite through the shortest path. Assuming that satellite i is the next hop of the optimal path selected by DSP algorithm, and satellite j is the next hop of the sub-optimal path, the congestion control is designed as follows:

Table 2. The ability of node to handle services

Data occupancy of node i	Congestion situation	Services that can be handled
$[0, AM_1]$	No	C1, C2, C3
$[AM_1, AM_2]$	Slight	C1, C2
$[AM_2, 1]$	Heavy	C1

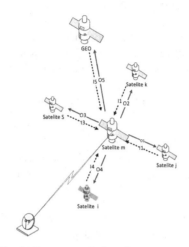

Fig. 5. Data transmission diagram of node

Step 1: Firstly, satellite m checks the congestion of all the neighbor satellites except satellite s. If all of them are in the congestion state, let packets C3 transmit through GEO satellite, and packets C1 and C2 transmit from link O3 to satellite s with re-selecting the path at satellite s. Otherwise go to the next step.

Step 2: Satellite m checks the congestion of the next hop satellite i which connected to the optimal path. If the packet occupancy of satellite i is $[0, AM_1]$, it means that this node is normal and no congestion occurs. At this time, all three types of data packets are sent to satellite i for transmission.

Step 3: If the packet occupancy of satellite i is $[AM_1, AM_2]$, it means that slight congestion occurs in this node. At this time, packets C1 and C2 are sent to satellite i, packets C3 are transmitted by satellite j, satellite k, or GEO satellite.

Step 4: If the packet occupancy of satellite i is $[AM_2, 1]$, it means that serious congestion occurs in this node. At this time, packet C1 will be sent to satellite i. Data packets of type C2 are transmitted by satellite j, while those of type C3 are transmitted by satellite k or GEO satellite.

3.2 Network Adaptation Model in the Case of Equipment Failure

The network is composed of nodes and links. Therefore, the failure of key nodes will lead complicated cascading effects, affecting the performance of the entire network. In this section, we first present an approach to evaluate the importance of node based on Node importance evaluation method based on transmission characteristics. Then, we analyze the failure of node and port in the network, and design an invulnerability routing scheme. Finally, we conduct a simulation to test the performance of this scheme.

(i) **Method for node importance evaluation**

The dynamic change of the satellite network makes the node dynamic as well. Firstly, we divide the whole simulation time into varying time slices, and obtain the importance of each node in different time slices. Then, we aggregate the node importance to obtain the total node importance throughout the transmission. And the node importance is determined by how many shortest paths the node undertakes in the transmission progress.

Definition 1: Distance between nodes H. We select the optimal path according to transmission delay. Usually, we consider the path that has the least hop count as the optimal path and regard hop count as communication distance. Let $H_{i,j}$ denote the hop count from node v_i to node v_j, and especially, when two nodes are unable to connected, we consider that $H_{i,j} = \infty$.

Definition 2: Node dependency F, which reflect the extent of the transmission progress depends on this node, is the ratio. The node dependency is the ratio of the number of paths passing through node i in all paths under the current transmission condition

$$F_i = \mu_1 \times \frac{P_{self}}{P_{total}} + \mu_2 \times \frac{P_{relay}}{P_{total}} \qquad (10)$$

where μ_1 and μ_2 are weights, P_{total} is the total number of paths in the network, P_{self} is the number of paths when node v_i is an accessing satellite, and P_{relay} is the number of paths when v_i is a relaying satellite.

Definition 3: Node importance I. In a single time slice, the importance of satellite nodes is affected by both the dependence of current node and the contribution of other nodes to the current node. And the contribution of other nodes to node v_i is mainly determined by the node dependence of them and the distance from the node v_i:

$$Con_{(j,i)} = \frac{F_j}{H_{i,j}}, i \neq j \qquad (11)$$

And the node dependence can be expressed by

$$I_i = F_j \times \sum_{j=1}^{N} Con_{(j,i)} \qquad (12)$$

where N is the number of satellite nodes.

The importance of a node throughout the transmission is the sum of each product of the node importance in each topological time slice and the weight of each time slice.

The weight of different time slice in the simulation can be calculated by

$$<w_1, w_2, \ldots, w_{n-1}, w_n> \ = \ \frac{T_1}{T} + \frac{T_2}{T} + \ldots + \frac{T_{n-1}}{T} + \frac{T_n}{T} \qquad (13)$$

In time T, the importance of satellite node i is

$$I_{i\_total} = w_1 \times I_{i\_1} + w_2 \times I_{i\_2} + \ldots + w_{n-1} \times I_{n\_1} + w_n \times I_n \qquad (14)$$

(ii) Network adaptation model in the case of node failure

When the satellite node fails, if the failure node is LEO node, then the four inter-satellite links connected with the LEO satellite will also fail. In order to ensure the instantaneous transmission of data, the entire network needs to obtain the information of the failed node in time, and then bypass the failure area and re-route. The area of the failed node is shown in Fig. 6.

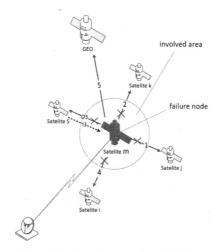

Fig. 6. Model diagram of node failure

The route update procedure when the node fails is as follows:

Step 1: The LEO satellite periodically sends packets to neighbors to inform them of their working conditions.

Step 2: When a LEO satellite finds that a neighbor node fails, it sends information about the failed satellite to the GEO management satellite.

Step 3: The GEO layer floods the message and sends invalidation information to the ground control center.

Step 4: After receiving the information, ground Control Center recalculates routes.

Step 5: The ground control center sends the calculation results to the LEO satellite.

Step 6: The LEO satellite re-transmits data based on the scheme in Sects. 3.1 and 3.2.

(iii) **Network adaptation model in the case of link failure**

When the network is transmitting data, a link connected to a satellite may fail due to external attack or damage caused by itself. The failure link is shown in Fig. 7. At this time, the satellite connected to the failure link immediately senses the situation and sends the service originally sent to the link to other paths connected with it. Suppose that link 4 is the optimal path selected by DSP algorithm, and link 1 is the sub-optimal path selected. Also suppose the maximum tolerance delay of packet C1 is t_{C1}, the length of optimal path is d_1, the length of sub-optimal path is d_2, and the data transmission rate of link is C. Therefore, when routing packets C1, the transmission conditions in the optimal path and the sub-optimal path are as follows (Table 3):

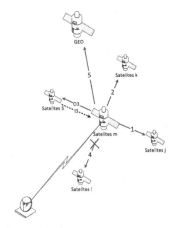

Fig. 7. Model diagram of link failure

Table 3. Transmission of optimal and sub-optimal paths

	$t_{C1} > \frac{d_2}{C}$	$\frac{d_1}{C} < t_{C1} < \frac{d_2}{C}$	$t_{C1} < \frac{d_1}{C}$
Optimal path	Can	Can	Cannot
Sub-optimal path	Can	Cannot	Cannot

In order to meet the minimum requirements for the delay of packets C1, the delay is at least greater than $\frac{d_1}{C}$. When satellite m receives the packets from satellite s, the steps of path selection are as follows:

Step 1: After receiving the packets from satellites, the satellite m first checks whether the links 1, 2, and 4 are valid. If the links 1, 2, and 4 are invalid, the satellite m sends the packets back to the satellite s through the link O3. When the satellite s checks that the packets is a packet sent by itself, the satellite m is marked as Unreachable, and then the packet is re-routed in the satellite s. Otherwise, do the following steps.

Algorithm1: invulnerability design process
Input: **packet$_{in}$**, time slice
Output: **packet$_{out}$**
1: While (TRUE)
2:     if (all ports are congested)
3:         discarded
4:     else
5:         if(occupancy is $[0, M_1]$)
6:             transmit all types packets
7:         else if(occupancy is $[M_1, M_2]$)
8:             transmit types C1 and C2
9:         else (occupancy is $[M_2, 1]$)
10:            transmit types C1
11:    if(all links are invalid)
12:        send back
13:    else
14:        if(optimal link is valid)
15:            transmit all types packets
16:        else
17:            transmit through sub-optimal
18:    if(next hop is failure)
20:            recalculates routes
21:    else
22:        if(occupancy is $[0, AM_1]$)
23:            transmit all types packets
24:        else if(occupancy is $[AM_1, AM_2]$)
25:            transmit types C1and C2
26:        else (occupancy is $[AM_2, 1]$)
27:            transmit types C1
28:    time slice++
29: end while

Fig. 8. Invulnerability design process

Step 2: The satellite m checks whether the optimal path 4 is invalid. If there is no failure, all types of packets are transmitted through the optimal path 4, otherwise, the sub-optimal path 1 is considered.

Step 3: Now, it is calculated whether t_{C1} is greater than $\frac{d_2}{C}$. If it is greater, all types of packets are transmitted through the sub-optimal path; otherwise, the packets C1 are directly discarded, and only packets C2 and C3 are transmitted.

The complete invulnerability design process of the space-based early warning system information network based on data transmission is shown in follow (Fig. 8):

4   Algorithm Simulation and Performance Analysis

4.1   Simulation Parameter Settings

OPNET simulation software is used to simulate the invulnerability routing algorithm of space-based warning system information network based on data transmission. The network layer model is shown in the figure below (Fig. 9):

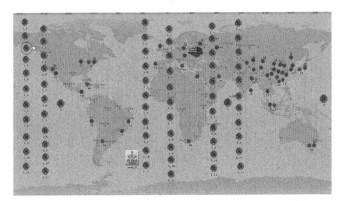

Fig. 9. OPNET network layer mode

Simulation time was set at 100 s and link bandwidth between LEO satellites was set at 20 Mbit/s; node cache size was set at 1 Mbit, and the cache size of each port is 250 kbit. The congestion threshold of both the port and the whole node is 0.8 times of the port and node size and the not congestion threshold is 0.5 times of the port and node cache and the ratio of packets C1, C2, and C3 is set to 1:3:6.

4.2   Simulation Result Analysis

In this paper, two sets of experiments are carried out. The first set of experiments was to verify the network performance changes under congestion. In order to verify the proposed routing strategy, this paper compares the DSP shortest path algorithm without congestion strategy and conducts experiments on port packet forwarding amount, packet loss rate and average end-to-end delay.

Figure 10 shows the packet forwarding amount of port connecting the optimal link and other ports under the DSP algorithm and routing policy proposed in this paper. As shown in the Fig. 11: the packet forwarding amount of the DSP algorithm connected to the optimal link increases slowly with the increase of the transmission rate. This is because the packet loss caused by congestion. And the packet forwarding amount of other ports under the DSP algorithm is 0. This is because the DSP algorithm does not perform traffic offloading. The port connecting the optimal link has a large drop in the number of packet forwarding under the routing strategy proposed in this paper, because traffic offloading is caused by congestion. Then the port only transmits packets C1, and the rest of the packets are transmitted through other ports.

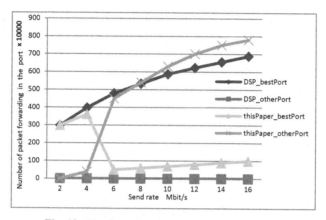

Fig. 10. Number of packets forwarding in the port

Figure 11 is a comparison between DSP algorithm and routing strategy proposed in this paper in terms of packet loss rate. At the beginning, there was no loss of packet in the two algorithms. With the increase of the sending rate, the node became congested. DSP algorithm only adopted the optimal path to transmit packets and did not offload the service. However, the routing strategy proposed in this paper makes different services

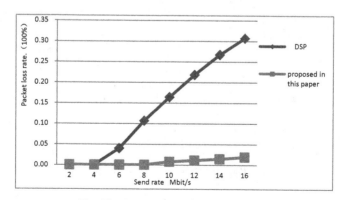

Fig. 11. Comparison of packet loss rates

transmit along different paths in case of node congestion, which not only guarantees the maximum non-loss of packets, but also meets the QoS requirements of different users.

Figure 12 is a comparison of transmission delay. Since the DSP algorithm does not divide the services, the transmission delay of all types of packets is the same. In the routing strategy proposed in this paper, the delay has been kept at a low level because packets C1 have the highest priority. Packets C2 need to be transmitted through the sub-optimal path when congestion occurs, so the transmission delay is low at the beginning, and then increases. Compared with packets C1 and C2, packets C3 have been in a state of high delay because they have no requirement for delay.

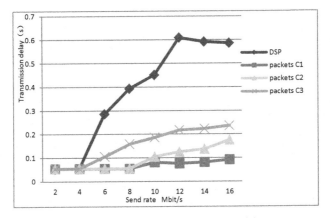

Fig. 12. Comparison of transmission delay

According to Fig. 13, when a link fail, the packet reception will drop briefly at beginning under the routing strategy proposed in this paper, then returns to the normal level,

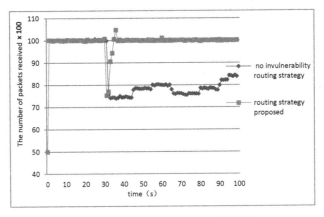

Fig. 13. Packets received in case of link failure

because the packet will choose the sub-optimal link for transmission. If the invulnerability routing strategy is not used, the amount of packets received after the link failure will be significantly reduced.

According to Fig. 14, when a node fails, the packet reception will drop briefly at the beginning under the routing strategy proposed in this paper, but then returns to the normal level, because the re-routing process of the satellite network. If the invulnerability routing strategy is not used, the number of packets received after the node failure will be significantly reduced.

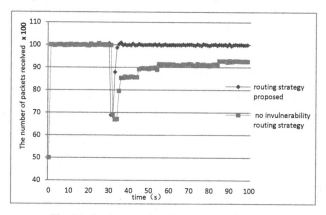

Fig. 14. Packets received in case of node failure

5 Conclusions

This paper takes the space-based early warning system information network as the research object, and studies the network invulnerability from two aspects. The first is the network invulnerability in the case of node congestion. The second is the network invulnerability in the case of equipment failure. At the same time, the network invulnerability in the case of node congestion can be divided into the in the case of single port congestion and all port congestion. In the design of network invulnerability under the condition of node congestion, the services are classified, and the congestion degree of ports and the whole node is divided into thresholds. Physical network vulnerability is divided into in the case of link failure and node failure. When the failure occurs, the whole nodes quickly perceive the failure node and carry out network reroute. Because of considering many kinds of vulnerability strategies, compared with the traditional DSP algorithm, the routing strategy proposed in this paper has good performance in packet loss rate, end-to-end delay and other aspects.

References

1. Li, H., Wu, Q., Xu, K., et al.: Progress and tendency of space and earth integrated network. Technol. Rev. **34**(14), 95–106 (2016)
2. Wang, J., Yang, J., et al.: Enlightenment of constructing integrated space-ground network information system. Command. Inf. Syst. Technol. **7**(04), 59–65 (2016)
3. He, X.: Study on satellite communication system's reliability based on invulnerability. University of Electronic Science and Technology (2013)
4. Newport, K.T., Schroeder, M.A., Whittaker, G.M.: Techniques for evaluating the nodal survivability of large networks. In: Military Communications Conference (1990)
5. Frank, H., Frisch, I.T.: Analysis and design of survivable networks. IEEE Trans. Commun. Technol. **18**(5), 501–519 (1970)
6. Barabasi, A.L., Albert, R.: Emergence of scaling in random networks. Science **286**, 509–512 (1999)
7. Huang, R., Li, W., et al.: Research on the invulnerability of combat SoS under different attack strategies. Command. Inf. Syst. Technol. **9**(03), 62–69 (2012)
8. Kawamoto, Y., Nishiyama, H., Kato, N., Kadowaki, N.: A traffic distribution technique to minimize packet delivery delay in multilayered satellite networks. IEEE Trans. Veh. Technol. **62**(7), 3315–3324 (2013)
9. Zhu, A.: Research on reliable transmission for inter-satellites multi-path based on optimization model. Nanjing University of Posts and Telecommunications (2016)
10. Nishiyama, H., Tada, Y., Kato, N., Yoshimura, N., Toyoshima, M., Kadowaki, N.: Toward optimized traffic distribution for efficient network capacity utilization in two-layered satellite networks. IEEE Trans. Veh. Technol. **62**(3), 1303–1313 (2013)
11. Yao, D., Wang, C., Shen, J.: Traffic-adaptive hybrid routing for double-layered. IEEE (2009)
12. Song, G., Chao, M., Yang, B., et al.: TLR: a traffic-light-based intelligent routing strategy for NGEO satellite IP networks. IEEE Trans. Wirel. Commun. **13**(6), 3380–3393 (2014)

Research on Space Information Network Protocol

Yongxue Yu[✉], Jiayu Xie, Yujue Wang, and Yin Zhou

Naval Research Academy, Beijing 100036, China
`yyxjuan@sina.com`

Abstract. This paper introduces four space information network protocol researched and applied internationally: space IP protocol, CCSDS protocol, protocol combining CCSDS and TCP/IP, and delay/interrupt tolerant network (DTN) protocol. The protocol architecture, protocol components and functions are introduced and analyzed. At the same time, the method of performance improvement of space satellite communication TCP protocol is briefly introduced.

Keywords: Space information network · CCSDS · DTN · TCP/IP

1 Introduction

The space information network is a network system that acquires, transmits and processes space information in real time on the basis of a space platform (such as synchronous satellite or medium or low orbit satellite, stratospheric balloon and manned or unmanned aerial vehicle). The space information network can serve major applications such as ocean sailing, emergency rescue, navigation and positioning, air transportation, and Aerospace measurement and control. It supports high dynamic, broadband real time transmission of ground observations downwards, and supports ultra long range, large delay reliable transmission of deep space detection upwards. Because the space communication environment has the characteristics of large link delay, high bit error rate, asymmetric forward reverse link and easy interruption, the existing Internet TCP/IP protocol cannot adapt well to this environment, CCSDS (Consultative Committee for Space Data Systems) has extended and improved the TCP/IP protocol, composed the SCPS protocol suite adapted to the space network. In order to adapt to the restricted network environment such as long delay and intermittent connection of space communication, a Delay/Disruption Tolerant Network (DTN) protocol is proposed.

2 Characteristics of Space Information Networks

Space information networks have many different characteristics compared to terrestrial networks and face the following challenges [1]:

(1) The space information network is sparse and the network topology changes frequently. The high speed operation of the satellite in orbit makes the network topology unstable.

© Springer Nature Singapore Pte Ltd. 2020
Q. Yu (Ed.): SINC 2019, CCIS 1169, pp. 163–171, 2020.
https://doi.org/10.1007/978-981-15-3442-3_13

(2) The orbits of satellites or spacecraft are pre-set, predictable, and very different from the two-dimensional motion models commonly used in mobile ad hoc networks.

(3) The link is asymmetric and the propagation delay is large. In order to reduce the overhead of terrestrial mobile terminals, many satellite systems employ a down-link/uplink asymmetric communication scheme. Due to the long distance of space data transmission, the transmission delay is much larger than that of the terrestrial network, and the asymmetry of the link makes it easy to reduce the efficiency of data transmission.

(4) The high error rate of data transmission link. The distance between the interstellar, inter-satellite, and star-to-earth links is large, and the data signal strength is proportionally attenuated as the transmission distance increases. At the same time, the transmission is easily distorted by various random factors such as sputum phenomenon, eclipse phenomenon, weather state (rain decay, etc.), shadow effect, multipath effect, ionization effect, etc., resulting in high data transmission error rate.

(5) Intermittent network connection. In a space network, due to the different orbits and speeds of the communication nodes, the stability of the communication link connection is poor, and the space environment is affected, and the link is interrupted frequently.

(6) The complexity of the network protocol. Spacecraft has a wide variety of functions. The application system is self-contained and the communication protocol standards are different. The terrestrial network protocol adapts to low latency data transmission. The long delay space network protocol does not match the terrestrial network protocol.

(7) The finiteness of node resources. The high cost and environmental constraints of spacecraft require spacecraft communication equipment to be small in size, light in weight, low in power consumption, and high in performance. The resource allocation of equipment is strictly limited, and the aerospace-grade anti-irradiation device is usually used on the star. To ensure high reliability, the performance of the device itself is much lower than that of terrestrial commercial devices, so the node resources are limited, and the calculation, storage, processing and transmission capabilities of the on-board load are relatively weak.

(8) The particularity of the space environment. The ionosphere, troposphere, and space electromagnetic activities easily destroy the communication function of the space network nodes. This requires strengthening the research on the node protection capability. The redundancy, reliability, and security of the system should be fully considered in the design.

3 Space Information Network Protocol System

At present, the space information network protocol system mainly includes: a space IP protocol system, a CCSDS protocol system, a protocol system combining CCSDS and TCP/IP, and a delay/interrupt tolerant network (DTN) protocol system. These protocol systems are interdependent. The mature technology of terrestrial Internet TCP/IP and the application verification of space IP provide a clear direction for the improvement of CCSDS recommendations. The protocol system combining CCSDS and IP can take advantage of TCP/IP. And can meet the requirements of space communication, DTN protocol architecture can solve the problem of reliable transmission under heterogeneous conditions in deep space environment.

3.1 Space IP Protocol System

The NASA Goddard Space Center established the OMNI project in 2002 to provide addressing capabilities, standard Internet protocols, and network application capabilities for future space missions, enabling users to establish end-to-end connectivity with their spacecraft anytime, anywhere. To this end, OMNI uses terrestrial commercial Internet technology for spacecraft, integrates space networks with terrestrial networks, and ensures that all network nodes operate as Internet nodes, enabling all-IP interconnection of terrestrial end users to spacecraft, reflecting different technological development ideas from CCSDS. The TCP/IP based protocol used by OMNI is shown in Fig. 1.

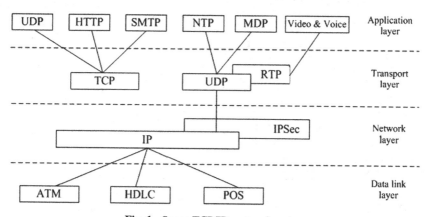

Fig. 1. Space TCP/IP protocol system

(1) Data link layer. Use data link layer protocols that are commercially available on the ground system, such as HDLC (Advanced Data Link Control), POS (Packet over SONET), and ATM (Asynchronous Transfer Mode). It is convenient for the air ground link and the commercial router to directly interface at the data link layer.

(2) Network layer. Use the IP protocol. Use IETF standard routing protocols, such as RIP, OSPF, BGP, etc. Use mobile IP protocol to solve the problem of single spacecraft flying over multiple ground stations. Use mobile network protocols (RFC3963) to solve the problem of multi access spacecraft moving through multiple ground stations.

(3) Transport layer. Use TCP and UDP protocols (and RTP protocol). The spacecraft real time telemetry data and payload data are transmitted using the UDP protocol. In the abnormal situation, the spacecraft is blindly controlled. Under normal circumstances, the spacecraft is controlled by the TCP protocol, and signaling information

related to the mobile IP protocol is transmitted. Real time multimedia communication is supported by the Real-time Transport Protocol (RTP). The RTP family consists of two protocols: Data Transfer Protocol (RTP) and Control Protocol (RTCP). RTP is responsible for transmitting data packets with real time information. It generally works on the base of UDP, it adds information such as time scale and serial number. The receiving end uses RTP to synchronize video and audio data. RTCP is responsible for monitoring network service quality, communication bandwidth, etc., and notifying it to the sender.

3.2 CCSDS Protocol System

The protocol architecture defined by CCSDS includes: application layer, transport layer, network layer, data link layer, and physical layer, each layer contains several combinable protocols. Figure 2 shows the CCSDS space communication protocol reference model.

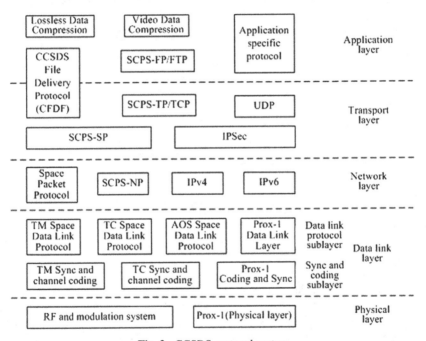

Fig. 2. CCSDS protocol system

(1) Physical layer. CCSDS defines a radio frequency and modulation system that specifies the physical layer protocol between the spacecraft and the ground monitoring station. The Prox-1 link protocol defines the physical layer characteristics of adjacent space links.
(2) Data link layer. The CCSDS Data Link sublayer defines the method of transmitting data over a space link using packets. The Synchronization and Channel Coding

sublayer defines a method of frame synchronizing and channel coding over the space link. The CCSDS Data Link Protocol sublayer consists of four protocols: TC Space Data Link Protocol, TM Space Data Link Protocol, AOS Space Data Link Protocol, and Data Link Layer of the Prox-1 Space Link Protocol. The above protocol specifies the ability to transmit data in a space link, collectively referred to as SDLP (space data link protocol). Correspondingly, CCSDS also specifies three specifications for the synchronization and channel coding sublayer of the data link layer: TM synchronization and channel coding, TC synchronization and channel coding, coding and synchronization layer protocol of the Prox-1 space link protocol.

(3) Network layer. CCSDS specifies two network layer protocols: Space Packet Protocol SPP, SCPS-NP (Space Communication Protocol Specification-Network Protocol), which implements the routing function of the space network. SPP is based on no connection and does not guarantee the sequential transmission and integrity of data. The core of the SPP is to configure the LDP (Logical Data Path) in advance, and use the Path ID instead of the complete end address to identify the LDP. This improves the efficiency of space information transmission, but is only suitable for static routing communication. Compared with the standard IP protocol, the SCPS-NP protocol has three improvements: NP provides four types of packet headers for users to choose between efficiency and function; both connection-oriented routing and connectionless routing are supported; compared with ICMP, SCPS Control Information Protocol (SCMP) provides a link break message. IPv4 and IPv6 packets of the Internet can also be transmitted over the space data link protocol, and can be reused with SPP, SCPS-NP or a space data link alone [3,4].

(4) Transport layer. The CCSDS transport protocol SCPS-TP provides end-to-end transport services to space communication users. The TP can identify and distinguish data due to data loss caused by network congestion, bit error or link interruption, and implement functions such as header compression, selective negative acknowledgment, time stamp, and rate control. The transport layer data is generally transmitted by the network layer protocol. In some cases, the transport layer data can also be directly transmitted by the link layer protocol. The TCP and UDP of the Internet can run on the IPv4, IPv6, and SCPS-NP. The SCPS Security Protocol (SCPS-SP) and Internet Security Protocol (IPSec) can be used with transport layer protocols to provide end-to-end data protection [2].

3.3   Protocol System Combining CCSDS and TCP/IP

In October 2000, the Jet Power Lab launched the Next Generation Space Internet (NGSI) project to study the use of the CCSDS protocol and the IP protocol to interconnect the space network with the terrestrial network to achieve a "space Internet" that supports future space missions. NGSI uses a combination of CCSDS and TCP/IP, which continues to use CCSDS recommendations at the data link layer, to use IP and its extension technologies at the network layer. The TCP/IP protocol and the CCSDS protocol, or other protocols suitable for space tasks, are used at the transport and application layers. The protocol system combining CCSDS and TCP/IP is shown in Fig. 3.

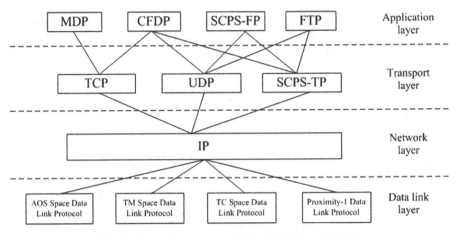

Fig. 3. Protocol system combining CCSDS and TCP/IP

3.4 Protocol System Based on Delay/Interruption Tolerant Network

Delay/Disruption Tolerant Network (DTN) is a new type of network system that can communicate in a restricted network environment such as long delay and intermittent connection. The concept of DTN was first proposed by the US Propulsion Laboratory in 2003. The Internet Research Task Force (IRTF) then formed the Delay/Interrupt Tolerant Network Research Group (DTNRG) on the DTN based on the Interstellar Network Research Group (INRG). In 2007, the DTN network architecture and the Bundle Protocol (BP) were proposed. In 2008, the convergence layer protocol was defined, including the TCPCLP (TCP Convergence Layer Protocol) protocol, the Saratoga protocol, and the

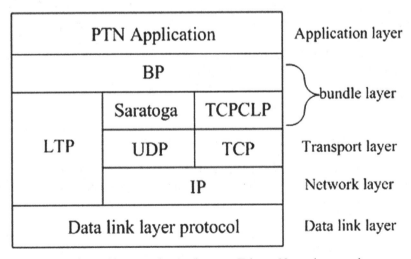

Fig. 4. Architecture of Delay/interrupt Tolerant Network protocol

Licklider Transmission Protocol (LTP). At the same time, in order to improve the short-comings of the BP protocol, a supplementary plan was formulated. Currently, researchers have conducted extensive research in the relevant fields of DTN and have carried out specific deployments and experiments. The core method of the delay/interrupt tolerant network architecture is to add an intermediate layer, the bundle layer, between the application layer and the lower layer (data link layer or transport layer). Figure 4 shows the architecture of the delay/interrupt tolerant network protocol [5].

4 TCP Protocol Performance Improvement Method of Space Communication

In a series of RFC documents developed by the Satellite Working Group and the Network Working Group in the Internet Engineering Task Force (IETF), there are many strategies for improving the performance of TCP protocols applied to satellite links. The solution considered from the TCP protocol itself is an end-to-end solution and a link layer solution. TCP enhancements in end-to-end solutions include: TCP enhancements (increase initial window, byte count, use larger window sizes, select acknowledgments, forward acknowledgments, etc.), TCP-Peach, and STCP (shared TCP). Link layer solutions include: ARQ protocol and adaptive forward error correction (FEC) [6]. These methods are all adapted to the space network environment by improving the TCP protocol itself, but they all have certain limitations. At the same time, there are special protocol clusters in the space communication environment to achieve better performance, such as the SCPS series protocol.

In order to make the space link part have higher transmission performance and ensure good data transmission capability under long delay and high bit error conditions, we can use the idea of "segmentation" from the perspective of the link itself. The link is "segmented" into segments, each segment using the relevant protocol applicable to that segment. There are two main technologies now, one is the Spoofing technology, and the other is the Splitting technology. Spoofing, as its name implies, is the meaning of protocol spoofing and is based on the network environment. Spoofing technology cuts long delay satellite links into segments and uses spoofing techniques to speed up link initiation. Since the Spoofing technology divides the entire TCP connection into segments, the protocol gateway is used to split the long-latency link in the entire link, and the TCP link is still used for data transmission in the satellite link part, and the gateway responds as a virtual destination node. The information is sent to the host at both ends, which speeds up the startup speed, and can also be applied to the retransmission of lost data, which greatly shortens the delay of the information at both ends of the transceiver, and improves the performance to some extent, because the transmission performance of the TCP protocol itself is limited by In the space communication environment, the improvement in overall performance is limited.

Another way to improve from the perspective of satellite links is TCP Splitting, which is to cut a complete link into segments and use the transmission protocol for the link itself in each link segment. The part of the link with high delay and high bit error in the entire link is separated from the whole. Splitting can fully adapt to network characteristics without modifying the client host and server protocols. As shown in Fig. 5, Splitting cuts it into three segments. The first segment is the client host to the protocol gateway. This part uses the original TCP protocol. The second segment is the protocol gateway to the satellite node. Use a proprietary protocol for the space link. The third segment is the satellite node to the satellite server side, which also uses the original TCP protocol. The most important feature of the Splitting technology is the end-to-end transmission. Compared to Spoofing, which increases the startup speed of link transmission and the speed of recovery from congestion, Splitting can greatly improve the transmission performance of complex space link segments.

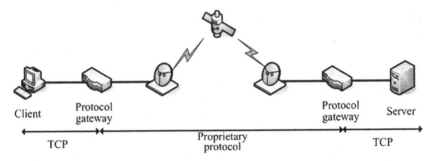

Fig. 5. Splitting technology link segment diagram

5 Conclusion

The four space information network protocol systems have their own characteristics. The advantage of the IP protocol system based on the terrestrial commercial standard is that the available software and hardware resources can be used to reduce the cost of the space mission. The interoperability between the space information network and the terrestrial network is better, but the processing capability of the spacecraft is also proposed. The CCSDS protocol system draws on the idea of terrestrial computer networks, and its design takes into account the new needs of future space missions. The CCSDS protocol performs better in terms of throughput and protocol overhead than the terrestrial network, but at the cost of certain CCSDS recommendations that do not fully conform to the layered principle of the ISO reference model and uses a different data format with the ground. The interoperability of the network is poor. Moreover, the CCSDS protocol does not solve the problem of dynamic routing. The protocol system combining CCSDS and TCP/IP not only utilizes the resources of commercial standards and good interoperability, but also has certain inheritance to existing CCSDS resources. The DTN-based protocol system addresses the deep space environment, and the deep space environment challenges the specific protocol performance of the data link layer to the transport layer. It needs to conduct in-depth research on layered protocols and cross-layer mechanisms, which will certainly promote the development of the DTN-based space IP protocol and the CCSDS protocol.

References

1. Zhou, J.: Research on key technologies of space integrated information network based on DTN. Doctoral Dissertation of Wuhan University, October 2013
2. Chen, Y., Meng, X.: Research and performance analysis of SCPS-TP protocol. Manned Space Flight **18**(1), 68–72 (2012)
3. Hao, X.: Research on space communication reliability transmission protocol. Master's thesis of Xidian University, March 2016
4. Zhang, M., Luo, G., Wang, J.: Research on reliable transmission protocol of space information network. J. Commun. **29**(6), 63–68 (2008)
5. Fall, K.: A delay-tolerant network architecture for challenged Internets. In: Proceedings of the ACM SIGCOMM, Karlsruhe, pp. 27–34. ACM Press (2003)
6. Gan, Z., Zhang, G.: New development of satellite communication technology. J. Commun. **27**(8), 2–9 (2006)

Theories and Methods of High Speed Transmission

Mutual Connection in 5G Based Space Information Networks: Opportunities and Challenges

Yuan Gao[1]([⊠]), Jiang Cao[1], Junsong Yin[1], Su Hu[2], Wanbin Tang[2], Xiangyang Li[1], and Tao Deng[3]

[1] Academy of Military Science of PLA, Beijing 100192, China
yuangao08@tsinghua.edu.cn
[2] University of Electronic Science and Technology of China, Chengdu 610054, China
[3] UNIT 78092, Chengdu 610000, China

Abstract. In recent years, satellite communication has becoming popular and develop very fast around the world. Recently, the fifth-generation mobile communication (5G) is about to enter the commercial use. As key characteristic of the 5G communication system, the integration of satellite communication and ground 5G has become a new hotspot. In this paper, we propose the summary of 5G in space based view, first of all, we introduce the status of 5G communication system, and its convergence to space wireless communication system. Then we analyze the development trends of satellite 5G systems, and then gives the initial idea of satellite 5G fusion, including architecture design, air interface design, based on SDN/NFV network virtualization deployment and protocol optimization, and discussed the possible problems of fusion; finally, the key technologies of satellite 5G fusion are sorted with opportunities and challenges.

Keywords: Mutual connection · Wireless fusion · 5G · Space information networks

1 Introduction

The main aspect of the satellite industry's important role in the 5G ecosystem is the ground segment. In fact, the VSAT platform is eager to adapt its architecture to 5G requirements [1]. The impact of 5G goes far beyond traditional satellite base station relay [2]. Therefore, all satellite communication applications, including emerging scenarios such as mobile, and future satellite functions such as video streaming multicast, will benefit from the adoption of the new 5G standard [3].

In depopulated areas, aircraft or ocean-going vessels that cannot be covered by ground 5G networks, satellites can provide economical and reliable network services that extend the network to places that are not reachable by terrestrial networks [4–6]. Satellite can provide continuous and uninterrupted network connection for IoT devices and mobile

This work is supported by National Nature Science Foundation of China (61701503).

carrier users such as airplanes, ships, trains, and automobiles. After the integration of satellite and 5G, the service capability of the 5G system can be greatly enhanced. Satellite's superior broadcast/multicast capabilities provide efficient data distribution services for network edges and user terminals [7].

In this paper, we discuss the opportunities and the challenges of 5G mutual connection wireless networks, the fusion will lead to new scenarios so that the future of space based wireless system is worthy to be discussed. The rest of this paper is organized as follows. In Sect.2, the Era of 5G and satellite communication system is given. In Sect. 3, we propose the summary of opportunities and challenges in mutual connection in 5G. Finally we give the conclusion of this talk.

2   The Era of 5 G

In recent years, satellite communication has ushered in a new wave of development around the world. The 5G on the ground is about to enter commercial use [8]. The integration of satellite communication and ground 5G has become a new hotspot in the field of satellite and ground.

2.1   5G System

In June 2018, with the freezing of the independent networking function of 5G New Air Interface, 5G has completed the first stage of comprehensive standardization [8] and entered into the stage of comprehensive industrialization. It is expected to be fully commercialized by 2020. Compared to previous mobile communication, the performance of 5G system is greatly improved, the peak rate can reach 10 Gbps in average and 20 Gbps peak, the commercial downlink for user rate can reach 100 Mbps to 1 Gbps [9], the connection density can reach 1 million per square kilometer, and the traffic density per square meter can be up to 10 Mbps, which can support high speed movement scenario under 500 km/h [10].

In terms of business capabilities, 5G can meet more abundant business needs. In the past 5 generations, mobile communication mainly focused on the communication between "people and people"; different in 5G, it is necessary to realize efficient communication between "people and things" and "objects and things", and finally realize "all things interconnection". The International Telecommunication Union (ITU) defines enhanced mobile broadband (eMBB), high-reliability low-latency communication (uRLLC), and large-scale machine communication (mMTC) as the three main scenarios for 5G [11–14].

5G could supports a wide range of operating bands including millimeter waves, by using high-frequency multi-antenna (Massive MIMO), efficient channel coding technology, non-orthogonal multiple access, multi-carrier and other key technologies to achieve higher spectral efficiency and system capacity; In the 5G network, the concept of the core network is linked to the edge of the network, thereby reducing the transmission delay of the data plane (D-Plane) and the control plane (C-PLANE), through software defined network (SDN)/network function virtualization (NFV) and other technologies [15, 16]. The control forwarding separation is realized, and the decoupling between the network element function and the physical entity is realized, thereby realizing efficient management and allocation of network resources.

In August 2018, in the third phase of China's 5G technology R&D trial organized by the IMT2020 promotion group, Huawei and Ericsson completed the C-band test of 5G independent networking (Standalone, SA), which indicates that 5G commercial has the foundation.. In the same period, ZTE and China Telecom realized the first 5G coverage and panoramic live broadcast test in Xiong'an District [18], demonstrating the application potential of 5G in the future of new smart cities and ecological governance. On the terminal side, Huawei released the commercial chip "Balong" in February 2018. It also supports Sub-6 GHz (low frequency) and mmWave (high frequency), supports SA and NSA, and can achieve a transmission rate of 2.3 Gbps.

2.2 Fusion of Satellite System

At the end of 2018, there were 1003 in-orbit communication satellites, accounting for 47% of the total number of satellites. Typical geosynchronous orbit satellite mobile communication systems include Inmarsat, Thuraya, TerreStar, and SkyTerra. The Inmarsat system adopts the Ka-band to realize the development from mobile communication to high-capacity and high-bandwidth. It can provide 50 Mbps reception and 5 Mbps transmission for 60 cm-diameter ground terminals. The US company's ViaSat is a typical broadband satellite communication system with a Viasat-1 capacity of 140 Gbps and a ViaSat-2 satellite capacity of 300 Gbps [19].

Typical low-orbit communication satellite constellations, such as the Iridium system, were proposed by Motorola in 1987. In 1998, the first generation system of 72 low-orbit communication satellite networks was completed, mainly for handheld mobile phone users. The global personal communication service, the next-generation system currently being deployed (Iridium Next), provides integrated services such as communications, climate change monitoring, and multi-spectral mapping [20].

In recent years, the development of Internet satellite constellations has grown by leaps and bounds. Its main features include: the use of medium and low orbits, compared to synchronous orbit satellites can greatly reduce the round-trip transmission delay, so that the experience of satellite transmission can be comparable with terrestrial fiber; using dozens or even hundreds of small satellite constellation networks Achieve large-scale coverage, greatly reduce satellite production costs through modular design, thereby reducing communication tariffs, providing users with affordable communication services; using Ka or Ku frequency bands, system capacity is greatly improved, for example, single beam of O3b can provide 1.6 Gbps transmission rate, 70 beams per star, single star of OneWeb will carry the capacity of 5–8 Gbps, and the total system capacity will exceed 7 Tbps, it can provide 50 Mbit/s Internet service for terminals with 0.36 m aperture antenna. It can also provide high-speed broadband Internet services in areas where traditional Internet installations are too expensive [21].

After several decades of development, China's satellite communication system has formed a certain scale of construction. High-throughput satellite communication systems based on fixed services and satellite mobile communication systems based on mobile services are currently being developed, and low-orbit communication satellites are also entering the experimental phase. In the field of civil satellite communications, the company mainly develops and develops the satellite and Asia-Pacific series of communication and broadcasting satellite systems. The communication services have basically achieved coverage in Asia, Europe, Africa and the Pacific, and ranked sixth in the global satellite space segment. At present, there are 15 civil communication satellites operating in the C, Ku and Ka bands in orbit. The development of high-throughput broadband satellites in China has just started.

In terms of low-orbit communication satellite systems, the systems being planned in China mainly include the National Science and Technology Innovation-2030 Major Project "The Low Earth Orbital Access Network of the Space-Ground Integrated Information Network", the "Hongyan Project" of the Aerospace Science and Technology Group and so on. The low-orbit access network has a track height of 800–1100 km, providing global seamless coverage of mobile and broadband communication services, supporting aviation/nautical surveillance, spectrum monitoring, navigation enhancement and wide-area Internet of Things (IoT) services. The geese constellation has a height of 1,100 km and consists of 324 satellites. It supports mobile communication, broadband Internet access, Internet of Things access, hotspot information push, navigation enhancement, and aviation navigation monitoring.

From the development history of satellite communication, the current satellite communication system is gradually developing towards the integration of heaven and ground heterogeneous networks together. On the one hand, the demand and market traction of the space-based network is ubiquitous, the advantages of space-based and terrestrial networks are combined and complemented, and various applications penetrate into every corner of the land, sea and sky and all aspects of people's lives. On the other hand, driven by scientific and technological innovation, the space-based network's capacity is rapidly increasing, the rate is significantly improved, services are continuously expanding, and costs are significantly reduced. It is subverting the traditional telecom industry concept and leading industrial innovation and business model innovation.

3   Opportunities and Challenges

The network architecture of the satellite-based convergence is based on the network architecture of the 5G mobile communication system. The satellite communication system is supported by a flexible access cloud, and the satellite signal processing is converted into standard IP and sent to the upper layer for unified processing. The forwarding cloud is based on Software Defined Network (SDN) technology, which implements control and forwarding separation, forwarding and service convergence. Control Cloud is based on Network Function Virtualization (NFV) implements the unique "network slicing" feature of 5G systems. By adding virtualized network element functions for satellite services and dynamically building new service chains to support satellite voice transmission, broadband access and Control of satellite communication scenarios such as base station backhaul.

3.1   Opportunities

Based on Software Defined Radio (SDR), future terminals will also have multi-standard connection and transmission capabilities, which can simultaneously support seamless switching of satellite networks and terrestrial mobile networks and even online. Based on the "three clouds" network architecture, combined with SDR, SDN, NFV and other technologies, the satellite-integrated network can achieve unified and coordinated management of satellite and terrestrial network resources, and based on separation of bearer and signaling, signaling and standards. Decoupling realizes unified control independent of access mode, achieves deep integration of satellite and terrestrial network, seamlessly combines effects, and provides users with transparent and consistent ubiquitous communication services.

Whether for satellite communications or terrestrial mobile communications systems, the lack of available spectrum has become an urgent problem to be solved. Due to the traditional spectrum authorization method of the ITU, in the satellite communication, the traditional L, S, C, Ku frequency bands have been fully saturated, and the competition for the Ka-band and even the millimeter-wave band has become fierce. The ground mobile communication system is not too much, 5G is used to meet the needs of the three major application scenarios of eMMB, uRLLC and mMTC. The frequency will cover the high, medium and low frequency bands, that is, the overall frequency band will be considered. The high frequency band is used to meet hotspot coverage and

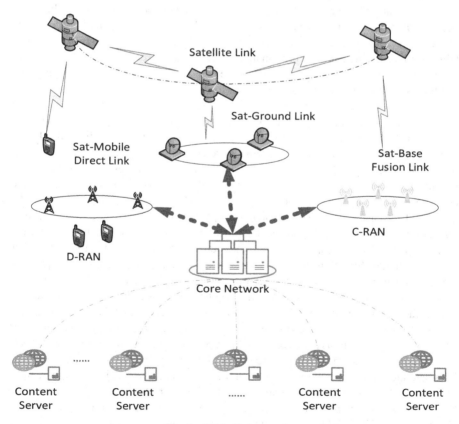

Fig. 1. 5G fusion networks

ultra-high-rate transmission, covering the available frequency bands of 24.25–86 GHz, and the mid-band of 3–6 GHz also overlaps with fixed-satellite and mobile services. The golden band of wireless communication below 3G also needs to be expanded, and frequency coordination problems arise with services such as satellite radio and television and satellite mobile communication. In the future, the separation between satellite communication and terrestrial mobile communication systems will be broken, and the mutual use of frequency resources by both parties will mean that the technical system, system components, and components and components will tend to be unified. Therefore, from the perspective of resource integration, the use of big data, cloud computing, artificial intelligence and other technologies to build a unified resource coordination platform to promote the sharing of frequency resources can provide a compatible basis for the deep integration of satellite communication systems and 5G systems (Fig. 1).

3.2 Challenges

The widely accepted challenges are mainly focused on LEO and GEO satellites.

The main problem of GEO satellites is that the transmission delay is large and the path loss is large. The transmission delay not only easily amplifies the impact of network congestion, but also makes it difficult for the control layer to perform on-time scheduling and coordination of communication resources. The 5G air interface technology described above often requires frequent interaction between the client and the control layer. The GEO satellite channel can be approximated as a constant reference channel. In fact, there is no need to adjust the resource allocation frequently according to changes in channel conditions. The control layer needs to adapt to the burstiness of the service. Then the access technology applicable to satellite needs to simplify the signaling process, realize the superposition of multi-user in the power domain through open-loop control, and adapt to the nonlinear changes brought by satellite forwarding. At the same time, the Hybrid Automatic Repeat request (HARQ) protocol used in 5G systems is prone to congestion in the transmission of satellite services. Therefore, it is also necessary to simplify the retransmission process and make more use of error correction codes or sources for the characteristics of long delay of satellite services. Fault tolerance technology reduces the number of retransmissions. The problem of path loss makes the user terminal have to maintain a certain transmission power, making it difficult for the device to miniaturize low power consumption. This is contrary to the practice of 5G reducing the transmit power of user terminals through ultra-density networking. Then, you can consider improving the transceiver capability of the terminal through a simple external device, or adopt a new large-diameter antenna such as a film antenna on the star to improve the satellite quality factor, and finally meet the miniaturization requirements of the user terminal.

The Low Earth Orbit (LEO) constellation can greatly reduce transmission delay and path loss, and can also increase system capacity through multiplexing. However, the high dynamic characteristics of satellites also cause Doppler shifts and complex inter-satellite routing problems. The goal of the 5G system is to provide communication services for high-mobility users with a speed of 500 km/h in the frequency band of 6 GHz. If the carrier frequency is 4 GHz, the system can adapt to a Doppler shift of about 1.9 kHz. The LEO satellite's flight speed is greater than 7 km/s. If the Ka-band is used, the Doppler shift will be greater than 400 kHz, which is much larger than the frequency offset range that the 5G system can adapt. Therefore, for the high dynamic characteristics of the satellite, it is necessary to optimize the air interface waveform, and make full use of the satellite ephemeris to compensate and correct the periodic Doppler shift. In addition, in the absence of a large number of ground station assistance, the routing problem of the inter-satellite link is also complicated, which may cause network congestion at the peak of the service and become a bottleneck problem. The routing algorithm of the inter-satellite link is closely related to the constellation design, the ground station planning, and the inter-satellite communication system. It is necessary to fully consider the appropriate routing mechanism at the beginning of the design.

4 Conclusion

In this paper, we discuss the fusion of space and ground segment in 5G wireless systems, the opportunities and challenges are discussed. The fusion of space-ground segment will help increase the coverage and the transmission speed.

References

1. Satellite role in 5G Eco-System & Spectrum identification for 5G some perspectives. 5G Radio Technology Seminar. Exploring Technical Challenges in the Emerging 5G Ecosystem, London, pp. 1–16 (2015)
2. Ge, C., et al.: QoE-assured live streaming via satellite backhaul in 5G networks. IEEE Trans. Broadcast. **65**(2), 381–391 (2019)
3. Son, H., Chong, Y.: Coexistence of 5G system with fixed satellite service earth station in the 3.8 GHz Band. In: 2018 International Conference on Information and Communication Technology Convergence (ICTC), Jeju, pp. 1070–1073 (2018)
4. Cassiau, N., Maret, L., Doré, J., Savin, V., Kténas, D.: Assessment of 5G NR physical layer for future satellite networks,. In: 2018 IEEE Global Conference on Signal and Information Processing (GlobalSIP), Anaheim, CA, USA, pp. 1020–1024 (2018)
5. Tan, H., Liu, Y., Feng, Z., Zhang, Q.: Coexistence analysis between 5G system and fixed-satellite service in 3400–3600 MHz. China Commun. **15**(11), 25–32 (2018)
6. Yuk, K., Branner, G.R., Cui, C.: Future directions for GaN in 5G and satellite communications. In: 2017 IEEE 60th International Midwest Symposium on Circuits and Systems (MWSCAS), Boston, MA, pp. 803–806 (2017)
7. Gineste, M., et al.: Narrowband IoT service provision to 5G user equipment via a satellite component. In: 2017 IEEE Globecom Workshops (GC Wkshps), Singapore, pp. 1–4 (2017)
8. Guidotti, A., et al.: Architectures and key technical challenges for 5G systems incorporating satellites. IEEE Trans. Veh. Technol. **68**(3), 2624–2639 (2019)
9. Araniti, G., Orsino, A., Cosmas, J., Molinaro, A., Iera, A.: A low computational-cost sub-grouping multicast scheme for emerging 5G-satellite networks. In: 2016 IEEE International Symposium on Broadband Multimedia Systems and Broadcasting (BMSB), Nara, pp. 1–6 (2016)
10. Araniti, G., Bisio, I., De Sanctis, M., Orsino, A., Cosmas, J.: Multimedia content delivery for emerging 5G-satellite networks. IEEE Trans. Broadcast. **62**(1), 10–23 (2016)
11. Yan, X., et al.: The application of power-domain non-orthogonal multiple access in satellite communication networks. IEEE Access **7**, 63531–63539 (2019)

12. Luglio, M., Romano, S.P., Roseti, C., Zampognaro, F.: Service delivery models for converged satellite-terrestrial 5 g network deployment: a satellite-assisted CDN use-case. IEEE Netw. **33**(1), 142–150 (2019)
13. Kourogiorgas, C., Papafragkakis, A. Z., Panagopoulos, A.D., Sakarellos, V.K.: Cooperative diversity performance of hybrid satellite and terrestrial millimeter wave backhaul 5G networks. In: 2017 International Workshop on Antenna Technology: Small Antennas, Innovative Structures, and Applications (iWAT), Athens, pp. 46–49 (2017)
14. Ramabadran, P., Madhuwantha, S., Afanasyev, P., Farrell, R., Dooley, J.: Wideband interleaved vector modulators for 5G wireless communications. In: 2018 IEEE MTT-S International Microwave Workshop Series on 5G Hardware and System Technologies (IMWS-5G), Dublin, pp. 1–3 (2018)
15. Artiga, X., Nunez-Martinez, J., Perez-Neira, A., Vela, G.J.L., Garcia, J.M.F., Ziaragkas, G.: Terrestrial-satellite integration in dynamic 5G backhaul networks. In: 2016 8th Advanced Satellite Multimedia Systems Conference and the 14th Signal Processing for Space Communications Workshop (ASMS/SPSC), Palma de Mallorca, pp. 1–6 (2016)
16. Giambene, G., Kota, S., Pillai, P.: Satellite-5G integration: a network perspective. IEEE Netw. **32**(5), 25–31 (2018)
17. Kodheli, O., Guidotti, A., Vanelli-Coralli, A.: Integration of satellites in 5G through LEO constellations. In: GLOBECOM 2017 - 2017 IEEE Global Communications Conference, Singapore, pp. 1–6 (2017)
18. Orsino, A., Araniti, G., Scopelliti, P., Gudkova, I.A., Samouylov, K.E., Iera, A.: Optimal subgroup configuration for multicast services over 5G-satellite systems. In: 2017 IEEE International Symposium on Broadband Multimedia Systems and Broadcasting (BMSB), Cagliari, pp. 1–6 (2017)
19. Evans, B.G.: The role of satellites in 5G. In: 2014 7th Advanced Satellite Multimedia Systems Conference and the 13th Signal Processing for Space Communications Workshop (ASMS/SPSC), Livorno, pp. 197–202 (2014)
20. Stevenson, R.A., Fotheringham, D., Freeman, T., Noel, T., Mason, T., Shafie, S.: High-throughput satellite connectivity for the constant contact vehicle. In: 2018 48th European Microwave Conference (EuMC), Madrid, pp. 316–319 (2018)
21. Watts, S., Aliu, O.G.: 5G resilient backhaul using integrated satellite networks. In: 2014 7th Advanced Satellite Multimedia Systems Conference and the 13th Signal Processing for Space Communications Workshop (ASMS/SPSC), Livorno, pp. 114–119 (2014)

Research on Inter-satellite Link Scheduling of GNSS Based on K-means Method

Tianyu Zhang[✉], Jianping Liu, Zhiyuan Li, and Jingwen Xu

State Key Laboratory of Astronautic Dynamics, Xi'an 710043, China
zhangtianyu@tangzip.com

Abstract. With the continuous deployment of Ka multi-beam equipment, the number of satellites on that can be managed by ground system simultaneously is increasing. For the navigation constellation, the number of optical nodes is increased, which further improves the navigation accuracy and time synchronization accuracy. More time slots are available to enable the navigation constellation to provide better extended service functions. However, the sharp increase in the number of visible satellites in China leads to more complex intersatellite visibility relations, which leads to the optimization of intersatellite link planning. In particular, the calculation amount of PDOP between satellites increases exponentially, and the traditional algorithm can no longer satisfy the real-time link planning. In the paper, the k-means method is used to replace the inter-satellite distance with the angle difference. By clustering constellation satellites, the fast calculation of inter-satellite PDOP value is realized and the inter-satellite link building algorithm is optimized, which provides a solution for the rapid inter-satellite link planning after the deployment of Ka multi-beam equipment.

Keywords: K-means · Inter-satellite link · PDOP · Link scheduling

1 Introduction

Global satellite navigation system realizes autonomous navigation and real-time transmission of constellation information by using inter satellite link. For this reason, the inter-satellite link constitutes a dynamic infinite network with precise measurement and data transmission functions between the stars and the earth of GNSS constellation [1–4]. The inter-satellite link is the core technical system of the global navigation system. It is an effective technical means to ensure system security and reliability, improve the overall performance of the system, and is an effective technical support for building a space-based information network. It is extremely important and has a wide range of applications in military and civilian fields. At the same time, the inter-satellite link is used for joint orbit determination to improve navigation accuracy. The realization of the above functions requires that the satellites in the entire constellation are fully connected through the wireless link, and the inter-satellite orbit determination also requires each satellite to have an orbital inter-satellite link [5–7].

When the global satellite navigation system runs off the ground for a certain number of days without an inter-satellite link, the user ranging accuracy (URA) will continue to

© Springer Nature Singapore Pte Ltd. 2020
Q. Yu (Ed.): SINC 2019, CCIS 1169, pp. 183–192, 2020.
https://doi.org/10.1007/978-981-15-3442-3_15

increase, and the user positioning accuracy will continue to increase, even becoming a kilometer level, which cannot meet the navigation requirements of the navigation system. Nor can it meet the global system security requirements [8]. Under the existing conditions, only by using the inter-satellite link, based on the high-precision measurement data of the inter-satellite, based on the high-precision measurement data between satellites, the satellite orbit clock difference parameters can be calculated autonomously, the ephemeris on the satellite can be updated and maintained autonomously, and the autonomous navigation independent of the ground operation control system can be realized, so as to ensure the safe and reliable operation of the global system.

2   Background

With the deployment of the Ka multi-beam antennas, the point-to-point measurement and control mode of the original single-beam antenna has been changed, and it has become a point to many measurement and control mode. This reduces the pressure on the use of ground-based payload management resources while increasing the number of satellites that can be directly linked to station resources. Based on this model, this paper adopts the method of all nodes selection for the inter-satellite link, and takes all the satellites that the ground station can build the link as nodes, which can effectively solve the problems of time slot scheduling and signal hops in the process of building the traditional inter-satellite link. In addition, it can obviously improve the autonomous navigation index and delay index in the inter-satellite link, and provide more free time for expanding services.

The highly dynamic network topology of inter-satellite links has always been an important difficulty in the inter-satellite link time-slot scheduling. Domestic and foreign scholars have proposed different solutions [9, 10]. Werner designed an inter-satellite link topology algorithm for the Earth's low-orbit satellite constellation. This algorithm can preferentially realize inter-satellite information transmission in constellation network information transmission [11]. Yang proposed a method of inter-satellite link scheduling based on clustering method, which can realize rapid inter-satellite chain building [12]. Based on the existing polling time-division duplex structure of GPS inter-satellite links available, Wang proposed an improved inter-satellite link scheduling method using a step-by-step scheduling method [13].

In this paper, according to the actual situation of the continuous deployment of multi-beam equipment, k-means method is used to cluster nodes according to the physical distance. In the slot scheduling of non nodes, by selecting nodes in different classes, the upper limit of PDOP value in slot scheduding is effectively reduced, a large amount of computation is avoided, and the overall planning efficiency is improved.

3   Related Theory

The inter-satellite link system consists of satellite system, payload management system, measurement and control system and other related business parts, inter-satellite link operation management system and extended users. In the planning of inter-satellite link building, it is necessary to meet the requirements of navigation constellation's own indicators and provide time slot interface for extended users as much as possible [14]. The full-node satellite selection strategy adopted in this paper satisfies the above two conditions well, but it will increase the number of node satellites and increase the PDOP value optimization calculation. In this paper, the time is discretized, the distribution of satellites in each time slice is analyzed, and the k-means method is used to cluster the satellites, so that the PDOP value constraint can be solved quickly.

3.1 The Process of Inter-satellite

In order to realize various requirements such as autonomous navigation of inter-satellite links and to complete various extension functions of inter-satellite links, it is necessary to plan the relationship between satellites. The chain-building scheduling process for inter-satellite links is as follows (Fig. 1):

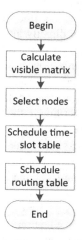

Fig. 1. Flowcharts of ISL

Calculation for the Visible Matrix. The visible matrix is a matrix that records whether satellites to satellites, satellites to ground stations can be directly linked in the constellation. The main factors considered in the visible model are: geometric visibility, signal reachability and antenna scanning range [15].

The simulation scenarios built using STK mainly include the Beidou medium-high-altitude satellite and the ground-measured operation and control node, and the visible performance analysis of the scene is required. Because the visible relationship between the satellites and the satellites is constantly changing, it is difficult to build chains. A more conventional method is to discretize the visible relationship between the satellites in the time dimension, and convert the dynamic changes into multiple static scenes for data analysis and processing. Under the condition of ensuring accuracy, the operation time cost is saved. In this paper, 10 min is used as a unit time slice for simulation analysis.

Selection for the Node Satellites. The node satellite is a transit station in the inter-satellite link, and the overseas satellite establishes a connection with the ground station, and is used for realizing the measurement and control command between the satellite overseas and the ground station, and relaying and transmitting the telemetry status information. The node satellite selection plays an important role in the time-slot scheduling of the inter-satellite link. The selection of the node satellite is related to the link allocation in the time-slot. According to the demand of the time delay indicator, the alien satellite needs to transmit the information to the ground within 3 s, and the alien satellite establishes the link with the node satellite at least every other time-slot, so there is a very high requirement for the number of node satellites. In this paper, relying on multi-beam equipment, a full-node satellite selection strategy is adopted, that is, the visible satellites of the ground station are set as node satellites.

Scheme for the Time-Slot Table. The time division system has flexible time-slot allocation capability. The time-slot of each node accessing the network are pre-allocated, synchronous access, and the allocation of access time-slots determines the connection status at any time in the entire constellation, directly affecting application services and network transmission performance [16], access time-slot allocation and finalization should be achieved through cross-layer optimization design. The chain-building relationship between satellites is calculated by the ground inter-satellite link management center and injected into the satellite in the form of a time-slot table. The satellite performs different satellite chain-building operations according to the received time-slot table.

The time-slot table is planned for each satellite's chain-building target and transmission and reception status in each time-slot, so that the GNSS constellation can ensure accurate data transmission and stable operation of the entire network at every moment. The indicators to be considered for time-slot table scheduling are time-delay indicators, autonomous navigation indicators, precision orbit determination indicators, and providing as many time-slot interfaces as possible for external services [17, 18]. For precision orbit determination and autonomous navigation, there are strict requirements for the PDOP value and the number of chains to be built. It is necessary to satisfy more than 9 chain-building numbers for any satellite, and the PDOP value is less than 1.5.

3.2 K-Means Clustering Algorithm

K-means clustering algorithm is a special case when solved by the maximum expectation algorithm of the Gaussian mixture model, in which the covariance of the normal distribution is a unit matrix and the posterior distribution of the hidden variables is a set of Dirac Delta functions. The k-means clustering algorithm, with its simplicity and efficiency, has become the most widely used algorithm in unsupervised learning such as clustering techniques.

Taking the Beidou global navigation constellation as an example, the principle of k-means clustering operation of the constellation satellite is as follows:

Input: constellation satellite set $D=\{x_1,x_2,\cdots,x_m\}$;

      Cluster Number $k=9$.

Process:

      Randomly selected samples from D as initial mean vectors. $\{\mu_1,\mu_2,\cdots,\mu_k\}$

      Repeat

          $C_i=\varnothing(1\le i\le k)$

          for $j=1,2,\cdots,m$ **do**

              calculate the distance between sample x_j and each mean vector μ_i

$(1\le i\le k)$: $d_{ji}=\left\|x_j-\mu_i\right\|$;

              According to the nearest mean vector to make the marker of x_j:

$\lambda_j=\arg\min_{i\in\{1,2,\dots,k\}} d_{ji}$;

              Delimit samples x_j into corresponding clusters: $C_{\lambda_j}=C_{\lambda_j}\cup\{x_j\}$;

          end
          for $i=1,2,\dots,k$

              calculate new mean vectors: $\mu'_i=\dfrac{1}{C_i}\sum_{x\in C_i}x$;

          if $\mu'_i\ne\mu_i$

              update current mean vectors μ_i as μ'_i

          else

              keep the current mean vector unchanged

          end

        end

      until current mean vectors are not updated.

    Output :clustering: $C=\{C_1,C_2,\cdots,C_k\}$

The k-means method is used to classify the domestic satellites into nine categories. According to the cluster correlation principle, the distance within the class is much smaller than the distance between the classes. Then one satellite from each of the nine categories can not only satisfy the condition that the number of links is greater than or equal to 9, but also make the distance between the selected satellites as large as possible, which excludes the case that the angle between the two selected satellites is too small.

3.3 PDOP

Position Dilution of Precision(PDOP) is a positional accuracy factor. The specific meaning is: because the quality of the observation results is related to the geometry of the satellite being measured and receiver and has a great influence on them, calculating the error caused by the above is called the intensity of accuracy. It is the strength of precision. The better the satellite distribution in the space, the higher the positioning accuracy (the smaller the value, the higher the accuracy). PDOP represents a parameter of the relationship between the three-dimensional positional positioning accuracy and the geometric configuration of the navigation platform.

As for the value of PDOP between satellites:

$$G_u = \begin{bmatrix} cos(el^1)sin(az^1) & cos(el^1)cos(az^1) & sin(el^1) & 1 \\ cos(el^2)sin(az^2) & cos(el^2)cos(az^2) & sin(el^2) & 1 \\ \vdots & \vdots & \vdots & \vdots \\ cos(el^k)sin(az^k) & cos(el^k)cos(az^k) & sin(el^k) & 1 \end{bmatrix} \tag{1}$$

where, $el^i = (i = 1, 2, \ldots, k)$ is the elevation angle of the i-th satellite, $az^i = (i = 1, 2, \ldots, k)$ is the azimuth of the i-th satellite.

geometric accuracy matrix is:

$$G = \left[G_u^T G_u \right]^{-1} = \begin{bmatrix} g_{11} & g_{12} & g_{13} & g_{14} \\ g_{21} & g_{22} & g_{23} & g_{24} \\ g_{31} & g_{32} & g_{33} & g_{34} \\ g_{41} & g_{42} & g_{43} & g_{44} \end{bmatrix} \tag{2}$$

three-dimensional PDOP is:

$$PDOP = (g_{11} + g_{22} + g_{33})^{1/2} \tag{3}$$

4 Simulation

4.1 Simulation Scenario

This paper uses STK to generate the corresponding orbital scene, extracts the orbital parameters, and uses MATLAB to calculate its visible matrix. The specific scenarios used in this article are as follows:

The space scene consists of 24 medium-orbit satellites, 3 high-orbit satellites and 3 geosynchronous orbit satellites. The 24 satellites form the Walker24/3/2 constellation with an orbital inclination of 55° and an eccentricity of 0; The orbital inclination of the high-orbit satellite is 55°, the longitude of the ascending node is 118°, and the phase difference is 120°; the three geosynchronous orbit satellites are 80°, 110.5° and 140°.

The ground scene consists of three ground stations, each equipped with multi-beam antenna equipment. It is assumed that the three ground stations are A in the northeast of China, B in the northwest, and C in Hainan.

The hardware and software environment used in the simulation scenario in this paper is shown in Table 1.

Table 1. Running environment configuration table

Hardware configuration	CPU	Intel(R) Core(TM) i7-7700 @3.60 GHz
	GPU	NVIDIA Quadro P2000
	RAM	32.0 GB
Software configuration	OS	Windows7
	Simulation system	STK10+MATLAB2010a

4.2 Process of Simulation

According to the latitude and longitude elevation information of the satellites and the ground stations of the navigation constellation, the relevant scene is constructed and the visible arc length is obtained. Taking 10 min as a unit time slice, the time is discretized and the visible matrix between satellites and between satellites and ground nodes is calculated. In the visible matrix, there are two states. Within 10 min, if the chain communication can be established all the time, the record is 1, otherwise 0.

The equipment used in the ground station in this paper is a Ka multi-beam antenna equipment, so it is assumed that satellites that can be seen with any ground station can be linked. According to the visible matrix between the satellites, the satellites that can be built chains with the ground stations are all set as node satellites. The number of optional node satellites is shown in Fig. 2. As can be seen from the figure, the number of node satellites is 11 to 17, accounting for about one-half of the total number of satellites.

It is necessary to schedule a time-slot table between the alien satellite and the domestic satellite to ensure the autonomous navigation and delay requirements of the entire constellation. Due to the large number of node satellites, the autonomous navigation and delay indicators are easy to satisfy, but the calculation of the entire time-slot scheduling is increased. The increase of the calculation amount is mainly reflected in how to obtain the optimal PDOP value. Using the k-means clustering method, a non-node satellite is selected, and the visible node satellites are clustered, and then the node satellites are selected in each class to build the chain, thereby completing the construction of the time-slot table quickly and accurately.

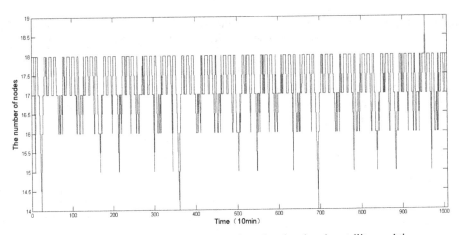

Fig. 2. The relationship between the number of optional node satellites and time

4.3 The Simulation Result

In the simulation, two methods are used to calculate the PDOP value, which are general method and K-means method. The specific reason why the ergodic method is not used in this paper is: in the time-slot scheduling, we need to compute a set of chain relationship combinations satisfying PDOP values for each satellite, while the traversal method is very time-consuming, and the average time for calculating the optimal chain relationships for each satellite is about 3575.5 s. The general method is to randomly select a group of satellites to calculate the PDOP value. In this paper, the random method is repeated 1000 times to ensure that several groups of chains can be selected to meet the PDOP value. The k-means method is described in Sect. 2.3.

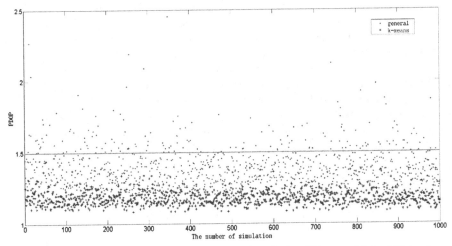

Fig. 3. Solving PDOP Value Result Graph by General Method and K-means Method (Color figure online)

The results in Fig. 3 can be obtained through experiments, in which the green line is the PDOP value indicator requirement, the red line is the PDOP value result of the general method in 1000 times, and the blue line is the PDOP of the k-means method in 1000 times. It can be seen from the figure that the PDOP value obtained by the random method is generally higher than that obtained by the k-means; the general method has a part of the PDOP value exceeding the requirement of 1.5.

In this paper, the specific data obtained by general method and K-means method are analyzed from three aspects: average PDOP value, maximum PDOP value and probability of meeting PDOP value requirement, as is shown in Table 2.

Table 2. Average computation time for each method

Method	Average value of PDOP	Max value of PDOP	Probability of satisfying PDOP index
k-means	1.1812	1.5323	99.8%
General	1.3880	2.4533	82.8%

As can be seen from the table, the k-means method is generally superior to the general method in terms of the average PDOP value and the maximum PDOP value. And the PDOP value obtained by the general method has certain risks, which can not meet the requirements of the inter satellite link building index, while the PDOP value obtained by the k-means method can basically meet the requirements of the inter satellite link index.

5 Conclusion

For the navigation constellation, the role of the inter-satellite link is constantly being explored. In the time-slot scheduling of the inter-satellite link, in addition to considering the requirements of the navigation constellation itself, it is also necessary to consider providing more time-slot interfaces for external services. With the continuous deployment of multi beam equipment, the number of visible satellites in China has increased, which effectively reduces the demand for inter satellite links for overseas data return and other services, but also increases the calculation of the optimal PDOP value. In this paper, by analyzing the physical distribution of constellation satellites in each time slice, the k-means method is used to cluster the space satellites, which effectively improves the efficiency of link planning between satellites, and provides technical support for the rapid realization of link planning of navigation constellation.

References

1. Jing, L., Ye, G.Q., Jiang, Z., et al.: A routing algorithm satisfied ground station distribution constraint for satellite constellation network. In: Science and Information Conference (SAI) (2015)

2. Wang, D.-x., Xin, J., Xue, F., et al.: Development and prospect of GNSS autonomous navigation based on inter-satellite link. J. Astronaut. **37**(11), 1279–1288 (2016)
3. Luo, D., Liu, Y., Liu, Y., et al.: Present status and development trends of inter-satellite link. Telecommun. Eng. **54**(7), 1016–1024 (2014)
4. Lin, Y.M., Qin, Z.Z., Chu, H.B., et al.: A satellite cross link-based GNSS distributed autonomous orbit determination algorithm. J. Astronaut. **31**(9), 2088–2094 (2010)
5. Li, Z., He, S., Liu, C., et al.: A topology design method of navigation satellite. Spacecr. Eng. **20**(3), 32–37 (2011)
6. Shi, L.Y., Wei, X., Tang, X.M.: A link assignment algorithm for GNSS with crosslink ranging. In: International Conference on Localization & GNSS (2011)
7. Yan, H., Guo, J., Wang, X., Zhang, Y., Sun, Y.: Topology analysis of inter-layer links for LEO/MEO double-layered satellite networks. In: Yu, Q. (ed.) SINC 2017. CCIS, vol. 803, pp. 145–158. Springer, Singapore (2018). https://doi.org/10.1007/978-981-10-7877-4_13
8. Huang, J., Su, Y., Liu, W., et al.: Optimization design of inter-satellite link (ISL) assignment parameters in GNSS based on genetic algorithm. Adv. Space Res. **60**(12), S0273117716307499 (2016)
9. SUN, J., Xu, B.: A kind of satellite link scheme for the navigation. In: Asia-pacific Conference on Information Theory (2010)
10. Cao, Y., Hu, X., Chen, J., et al.: Initial analysis of the BDS satellite autonomous integrity monitoring capability. GPS Solut. **23**(2), 35 (2019)
11. Werner, M.: Topological design, routing and capacity dimensioning for ISL network in broadband Leo satellite system. Int. J. Satell. Commun. **19**(6), 499–527 (2011)
12. Yang, D.N., Yang, J., Xu, P.J.: Timeslot scheduling of inter-satellite links based on a system of a narrow beam with time division. GPS Solut. **21**(3), 999–1011 (2017)
13. Wang, H., Chen, Q., Jia, W., Tang, C.: Research on autonomous orbit determination test based on BDS inter-satellite-link on-orbit data. In: Sun, J., Liu, J., Yang, Y., Fan, S., Yu, W. (eds.) CSNC 2017. LNEE, vol. 439, pp. 89–99. Springer, Singapore (2017). https://doi.org/10.1007/978-981-10-4594-3_9
14. Zhang, T.-y., Ye, G.-q., Li, J., Xu, J.-w.: FTS-based link assignment and routing in GNSS constellation network. In: Sun, J., Liu, J., Yang, Y., Fan, S., Yu, W. (eds.) CSNC 2017. LNEE, vol. 438, pp. 155–165. Springer, Singapore (2017). https://doi.org/10.1007/978-981-10-4591-2_13
15. Shi, L.Y., Xiang, W., Tang, X.M.: A link assignment algorithm for GNSS with crosslink ranging. In: International Conference on Localization & GNSS (2011)
16. Niu, F., Han, C., Zhang, Y.: Design and simulation for satellite autonomous integrity monitoring based on inter-satellite-links. Acta Geodaetica Cartogr. Sin. **40**(S1), 73–79 (2011)
17. Yong, X., Chang, Q.: On new measurement and communication techniques of GNSS inter-satellite links. Sci. China **55**(1), 285–294 (2012)
18. Wang, D.X., Xin, J., Xue, F., et al.: Development and prospect of GNSS autonomous navigation based on inter-satellite link. J. Astronaut. **37**, 1279–1285 (2016)

Research and Analysis of Node Satellites Selection Strategy Based on Navigation System

Zhiyuan Li[1,2(✉)] (iD), Tianyu Zhang[1,2], Jianping Liu[1,2], Ming Wang[2], and Jian Zhang[2]

[1] State Key Laboratory of Astronautic Dynamics, Xi'an 710043, China
med05@163.com
[2] Xi'an Satellite Control Center, Xi'an 710043, China

Abstract. In view of the large number of in-orbit satellites currently in China and the shortage of ground TT&C and control resources, combining with the networking requirements of China's navigation constellation and the current situation and development of the ground station network, this paper proposes two inter-satellite strategies based on the various service requirements of the navigation system, and carries out the utilization of ground resources separately. Through simulation analysis, the validity of the strategy is verified, and the optimal deployment schedule of ground resources is proposed, which provides effective assistant decision for the optimization of the network scheme and satellites management next.

Keywords: GNSS constellation · Inter-satellite link · Node satellites selection · Ground resources

1 Introduction

The Global Navigation Satellite System (GNSS) provides multi-functional services such as navigation, positioning and timing for global users in the information age, and plays an irreplaceable role in the country's economic development and national defense construction. At present, there are four major global satellite navigation systems in the world, which are named Global Positioning System (GPS) in the United States, Global Navigation Satellite System (GLONASS) in Russia, the Galileo System in Europe and COMPASS Navigation Satellite System in China. The navigation satellite system can build a dynamic space-ground wireless network with precise measurement and data transmission between the global satellite system and ground through inter-satellite links. It can realize autonomous navigation and efficient transmission of instruction, information and data, reduce the load on the main station and the dependence on the ground station, and improve the autonomy, robustness of the GNSS constellation network, as well as the real-time and accuracy of the data transmission [1]. China's COMPASS Navigation Satellite System officially opened basic navigation services in December 2018, and plans to implement a complete system networking in April 2020.

In the world aerospace field, the United States and Russia can deploy ground stations globally, and spacecraft can be measured and controlled at full time. As a result of many

This work was supported in part by the Nature Science Foundation of China under Grant 91638202.

© Springer Nature Singapore Pte Ltd. 2020
Q. Yu (Ed.): SINC 2019, CCIS 1169, pp. 193–206, 2020.
https://doi.org/10.1007/978-981-15-3442-3_16

factors, such as national boundaries and historical factors, China's telemetry, track and command (TT&C) and management network coverage rate is less than one-fifth among these countries. Now, most of the TT&C and payload-management ground stations of China owned or leased are distributed home. The number of overseas ground stations is very limited. The frequency covered by ground stations is mainly S-band single-beam, and some payload-management stations are deployed with L-band or Ka-band equipment. With the rapid development of the aerospace industry, the number of in-orbit satellites in China has increased dramatically, and the ground resources available have become scarce. How to make full use of the inter-satellite link technology, satisfy the GNSS constellation autonomous navigation and delay requirements as precondition, using less ground resources has great values both in practice and application. At present, a number of studies have been carried out based on inter-satellite links, which are oriented to the time synchronization, precise orbit determination and routing strategy of the navigation constellation [2–4]. However, after satisfying the system performance mentioned above, how to use the ground resources effectively and optimize the strategies of utilization of them badly, there are few related studies. Based on the current situation of China's navigation system and the requirements of networking, this paper analyzes the use of ground resources by inter-satellite links and carries out two strategies for using ground resources. Through simulation, the optimal strategy for maximizing the utilization of ground resources is proposed, which provided assistant decision for the GNSS network scheme and satellites management next.

2   Inter-satellite Link System Scheduling for GNSS

China's monitoring ground station for navigation system is mainly deployed home, and the coverage of MEO satellite observation arc is only 34%. In the absence of inter-satellite links, the space signal of user ranging accuracy(URA) of the MEO satellite is about 2 meters, which cannot meet the requirements of the global system design performance of URA 0.5 m, and also cannot achieve the performance of global system service [5]. Under the current conditions, only inter-satellite links can be used for joint precise orbit determination between space-ground and time synchronization and for realization of high-precision satellite orbit and clock correction calculation, so as to ensure the realization of service performance of global systems [6, 7].

By using inter-satellite links, the navigation system can realize the relaying distribution and return of TT&C and telemetry status data in entire constellation without stations overseas, improve the real-time performance of status monitoring, and give full play to its advantages in dealing with sudden failures, so as to improve the control and management level of satellites overseas. The composition of the inter-satellite link system of the navigation system is shown in Fig. 1.

In order to realize the autonomous navigation of inter-satellite links and other requirements, as well as to complete the expansion functions of inter-satellite links, it is necessary to schedule the timeslot relation of inter-satellite and space-ground links. Link scheduling process for inter-satellite links follows these steps as below.

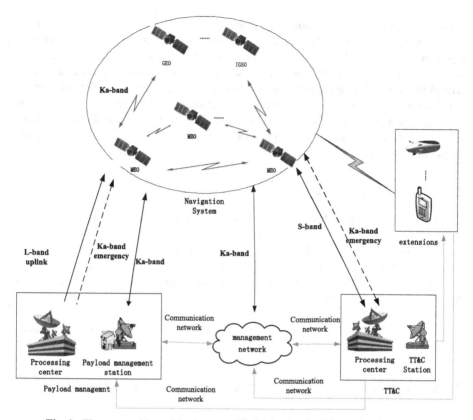

Fig. 1. The composition of the inter-satellite link system of the navigation system

2.1 Calculating the Visible Matrix

The visible matrix is a matrix that records whether satellites and satellites, satellites and ground stations can be directly linked in the constellation. The main factors that can be seen in the visible model are: geometric visibility, signal reachability, and antenna scan range [8, 9].

After using STK to build a simulation scene containing medium and high orbit satellites, TT&C and payload-management ground stations, it is necessary to make visible performance analysis. The main method is to discretize the visible relationship between satellites and satellites in the time dimension for data analysis and processing. Under the condition of ensuring accuracy, the cost of operation time is saved. In this paper, 10 min is used as a unit time slice for simulation analysis.

2.2 Selecting Node Satellites

Since China's TT&C and payload-management gound stations are mostly distributed home, they cannot be built globally. Therefore, data transmission between visible satellites home and invisible satellites abroad is a problem that needs to be addressed. We refer to the satellites visible to the ground station as domestic satellites, and the satellites invisible as overseas satellites. The TT&C or payload-management ground station want to inject remote command data into the overseas satellites, and the overseas satellites want to send or reply remote data to the ground station, they both need "a relay node". This relay node, which is a satellite that is visible to the ground station and the target overseas satellites at the same time, is defined as a node satellite. The node satellites selection plays an important role in the timeslot scheduling of the inter-satellite link. The selection of the node satellites is related to the link allocation in the timeslot and also the utilization of ground resources.

2.3 Scheduling Timeslot Table

The time division system has flexible timeslot allocation capability. The timeslots of each node accessing the network are pre-allocated [10], accessed synchronously, and the allocation of access timeslots determines the connection status at any time in the entire constellation, directly affecting application services and network transmission performance. Access slot allocation and final determination should be achieved through cross-layer optimization design [11, 12].

The timeslot table is scheduled for each satellite's link-building target and transmission and reception status in each timeslot, so that the GNSS constellation can ensure accurate data transmission and stable operation of the entire network at any time.

3 Analysis of the Ground Resources by Inter-satellite Links

3.1 Introduction of TT&C and Payload-Management Station

The TT&C station mainly completes tasks such as the bet of the satellite remote command data and the distribution and return of the telemetry data. In this paper, four ground stations are selected in the eastern, western, southern and central parts of China to form a measurement and control ground network supporting the measurement and control service of GNSS constellation. Four ground stations are deployed in location A, B, C, and D respectively. Among them, B station and C station deploy 16-beam multi-beam antennas, and A station and D station deploy several sets of single-beam antennas. All ground station equipment operates in the S-band and can establish S-band links with satellites. Each ground station is under the unified dispatch of the Xi'an Center to jointly complete the measurement and control tasks. In the case of an emergency, the Ka-equipment of the Ka-management station is used for emergency measurement and control.

The payload-management station mainly establishes the interaction of the payload data with the satellite. The payload-management data mainly refers to some functional and operational data such as satellite orbit determination and system differential calculation. In order to support the GNSS business, this paper selects three ground payload-management stations, which are deployed in Beijing, location A and location C respectively. Among them, Beijing deployed 16-beam equipment for the main payload-management station. From the perspective of the number of injection station beams, it has the ability to inject all visible satellites over the territory. The global system adopts the uplink injection mode of domestic station and the inter-satellite link in combination, which mainly relies on the ground calculation equipment and injection station in China, builds the link to the node satellites, and realizes the uplink data injection of the whole network satellite through secondary distribution of inter-satellite link.

3.2 Analysis of Node Satellites' Performance

During the operation of inter-satellite link system, the support of ground resources is needed, which is mainly reflected in the communication between ground node and space node which is called node satellite [13, 14]. According to the analysis of inter-satellite visibility, the average total number of MEO satellites is about 20, excluding the invisible MEO satellites and the visible GEO/IGSO high-orbit satellites currently. GNSS constellation uses a single frequency, which operates in time-division multiplexing half-duplex system. We set the timeslot unit of the time division system as 1.5 s, using half-duplex working mode, then build the two-satellites link every 3 s the average super frame of each MEO satellite is set to 60 s, we called 60 s a time cycle.

For the technical requirements of inter-satellite link measurement, the main indicators related to the schedule of the timeslot table are the minimum number of constellation links and the set distribution of single-satellite pseudo-range measurements, which is measured by the value of position dilution of precision (PDOP) [15]. Specifications are shown in Table 1.

Table 1. The technical requirements of inter-satellite link measurement

Main indicators	Precise orbit determination	Autonomous navigation	Time system
Minimum number of constellation links	≥ 6	≥ 9	≥ 3
PDOP	<3	<1.5	—

According to GNSS constellation autonomous navigation positioning accuracy requirements, PDOP value should be less than 1.5. From the table above, we can see that the minimum link number of the constellation is 9 at least, and the distribution of single satellites pseudo-range measurement set satisfies PDOP < 1.5. According to the inter-satellite link time-division system, it is necessary to build links with at least nine different constellation satellites during 20 slots of each time cycle. For the value of PDOP, the more links the same satellite has built in 60 s, the smaller the value of PDOP is.

The schematic diagram of the inter-satellite link is shown in Fig. 2. In terms of inter-satellite link measurement technology, transmission delay requirements and idle slots provided for extended services, the more node satellites, the better. However, from the point of view of saving ground resources, the fewer node satellites are, the better [16]. Therefore, how to minimize the number of node satellites is a trade-off issue under the premise of ensuring the instructional requirements.

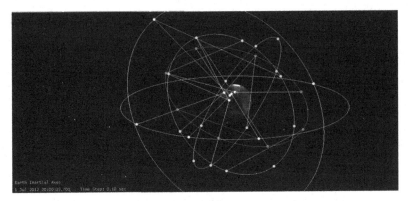

Fig. 2. The schematic diagram of the inter-satellite link

3.3 Analysis on Strategic Choice

To solve the above problems, two node satellite selection strategies are proposed in this paper. In the case of maximizing the use of GNSS constellation, in addition to satisfying the autonomous navigation and delay requirements of navigation constellation itself, more timeslot interfaces are needed for extended services [17], so maximizing node satellite selection strategy can be adopted; while considering how to meet the autonomous navigation and delay requirements of navigation constellation but reduce the number of ground resources, the strategy of minimizing the use of ground resources should be adopted. In order to facilitate qualitative analysis, the former may consider all visible satellites to ground stations as node satellites, which is called full-node-satellites selection strategy, while the latter aims at reducing the utilization of ground resources, which is called min-node-satellites selection strategy.

In the full-node-satellites selection strategy, satellites available in the territory can be selected as node satellites. For the TT&C station, the deployed multi-beam antenna is preferred, and the single-beam equipment is supplemented when the multi-beam antenna cannot meet the requirements of the node satellites for measurement and control; and as for the payload-management station, the main station Beijing is preferred, supplemented by other stations. For the GNSS constellation, 7 days is a regression period. Under the conditions of the above strategy, the coverage rate of the entire node satellites and the completion of the full-time task can be quantified from the two aspects of TT&C resources and payload resources in a regression period to determine the effectiveness between two strategies.

In the min-node-satellites selection strategy, firstly, as few ground stations as possible should be used, and the nodes of the multi-beam equipment are used preferentially, by increasing the number of node satellites and ground stations continuously and recursively, analyzing the requirements of the navigation system, min-node-satellites selection strategy is proposed. In order to save the use of ground resources, the TT&C resources can give priority to the TT&C stations of multi-beam equipment, and the payload-management resources can give priority to Beijing station as the main payload-management station. Under the policy conditions mentioned above, three indicators, such as transmission delay, PDOP value and the number of link-building, can be used as constraints to determine the number of node satellites to be selected.

3.4 Modeling Analysis

For the command data uplink injection of overseas satellites, the reception of downlink telemetry data and the return of other transaction data, it is necessary to establish an inter-satellite link with the node-satellite and forward it to the ground. For full-node-satellites selection strategy, the available satellites visible are selected as node-satellites. For the min-node-satellites selection strategy, it is necessary to establish a visible relationship model between node-satellite and non-node-satellite to determine the optimal selection. We set the scanning range of antenna for high-orbit satellites to be 45° and medium-orbit satellites to be 60°.

The inter-satellite geometry and the scanning range of inter-satellite antennas need to be considered in the inter-satellite model. The basic conditions for inter-satellite visibility are as follows [15].

$$r_{os} > h \tag{1}$$

$$L_{ij} \leq \min\{L_g, L_s\} \tag{2}$$

$$\alpha_{ij} < \alpha_i \, (\gamma_{ij} \leq \beta) \text{ or}$$
$$90° - \beta < \alpha_{ij} < \alpha_i (\gamma_{ij} > \beta) \tag{3}$$

Where r_{os} is the distance between geocentric and satellite links; h is the distance between the earth's core and the top of the ionosphere; L_{ij} is the distance between satellites I and j; L_g is upper limit of geometric visible distance between satellites; L_s is the upper limit of distance intersatellite signal can reach; α_{ij} is coverage angle of inter-satellite antenna; α_i is half power angle of inter-satellite antenna beam; γ_{ij} is the geocentric angle between satellite I and j; β is the maximum coverage angle of the inter-satellite antenna to the earth;

After obtaining the visibility matrix between the satellite and the ground, the optimal visible satellite is selected as the node-satellite based on the visible time of the ground station to the visible satellite and the number of the satellites when the node-satellite is chosen.

Inter-satellite visibility can be divided into persistent and non-persistent visibility. Under GNSS constellation time-division system, each MEO is linked with 20 satellites. For each satellite, 8 of all visible satellites are continuously visible and 12 are intermittently visible. Continuous visible link topology is relatively fixed, DPOP value changes relatively small and link selectivity is weak, but it can lay a foundation for link scheduling to measure communication performance. On the contrary, the non-persistent visible link topology changes regularly, DPOP values change greatly, and link selectivity is strong, which can optimize link performance. Therefore, according to the requirement of inter-satellite link-building requirements, continuous visible links can be used as the basis of link scheduling in the cycle, and non-continuous visible links can be used to improve the performance of link scheduling.

During the link scheduling, on the basis of following the scheduled rules proposed above, link-building targets should be planned for each satellite as much as possible, and meet the traffic of all satellites in the link for communication. In a finite timeslot, for any satellite between i and j, the constraints of link scheduling per timeslot are as follows.

(a) The number of links established from a satellite should not exceed the total number of links built.

$$\sum_j V_{ij} u_{ij} \leq N_t \tag{4}$$

(b) The connectivity from satellite i to satellite j is the same as that from satellite j to satellite i.

$$u_{ij} = u_{ji} \tag{5}$$

(c) The communication traffic between satellite i and satellite j equals the sum of the satellite traffic passing between them.

$$f_{ij} = \sum_m \sum_n x_{ij}^{mn} \tag{6}$$

(d) The communication between satellite i and satellite j in limited timeslot cannot be more than that in the whole period of GNSS constellation.

$$f_{ij} \leq C * u_{ij} \tag{7}$$

The initial satellite's traffic should be equal to that received by the terminal satellite. The total amount of information received and received in an inter-satellite link is the same.

$$\sum_j x_{ji}^{mn} - \sum_j x_{ij}^{mn} = \begin{cases} -O^{mn} & if\,i = m \\ O^{mn} & if\,i = n \\ 0 & otherwise \end{cases} \quad \forall m, n \; and \; x_{ij}^{mn} \geq 0 \tag{8}$$

Where u_{ij} is the connectivity between satellite i and satellite j $u_{ij} = 1$ means link, $u_{ij} = 0$ means unlink); V_{ij} is the visibility of satellite i and j (1 means visibility, 0 means invisibility); N_t is the maximum number of links building in the t-th finite slot; O^{mn} is the traffic needed from satellite m to satellite n; x_{ij}^{mn} is the traffic between satellite m and satellite n during the transmission from satellite i to satellite j; f_{ij} is the total traffic from satellite i to satellite j.

Using the link scheduling model mentioned above, the selection range of node satellites can be basically determined. In addition, considering the unified TT&C and payload-management transaction, the low-latitude ground station has the best field of view home. Therefore, the satellite with long visibility in every pass to the low-latitude ground station should be given priority as the node-satellite.

4   Simulation Analysis

This paper uses STK to simulate GNSS constellation scene for verification. The navigation system satellite is configured as "3GEO+3IGSO+24MEO" with a total of 30 satellites, of which the MEO satellite adopts the Walker 24/3/1 constellation configuration.

4.1   Full-Node Satellites Selection Strategy

Analysis on TT&C Resources

According to the selection criteria of ground stations for full-node satellite selection strategy, B and C stations are preferred. For satellites invisible to B and C, the TT&C equipment of A and D is used. Through simulation, the fluctuation figure of the number of tracking satellites of each station can be obtained, as shown in Fig. 3. As can be seen from the figure, A station and B station work continuously in seven days. The number of tracking satellites in each timeslot is about 8, and the maximum number of tracking satellites in single station and single equipment is 12. Two ground stations can fulfill the above tracking requirements. According to the analysis of single time point, for example, at 780 min, four satellites are only visible to A. A station needs at least four single-beam antennas to meet the normal operation of the whole system.

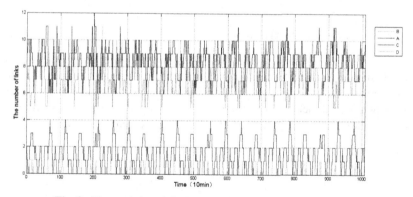

Fig. 3. The statistics of the number of links for the TT&C station

Due to geographical location, the amount of A's TT&C tasks will be higher than that of D's. The tracking time of A station during 7 days is 77.88%, and that of D is 12.10%. For the entire TT&C network, the coverage of the entire node satellite within 100 days of the four TT&C ground stations is 100%, and the task of measuring and controlling the entire node satellites can be completed at all times.

Analysis on Payload Resources

According to the scheduling principle, the priority use of the payload-management station is in Beijing, and supplemented by the others, the statistics of the number of links for the TT&C station is obtained, as shown in Fig. 4. As can be seen from the figure, the number of node satellites directly connected with Beijing station is basically maintained at 11 to 16, but at 9940 min, the number of links is 17, exceeding the upper limit of 16 beams for Beijing station, so it needs to be shunted to A or C station. For the two payload-management stations in A and C, as the auxiliary station, the equipment available can meet the requirements.

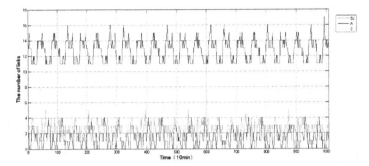

Fig. 4. The statistics of the number of links for the payload-management station

For the entire network, three payload-management stations including multi-beam equipment in Beijing, A and C, its coverage rate is 95.99% for the entire node in 7 days, and basically complete the tasks for the entire node satellites in full time.

4.2 Min-Node-Satellites Selection Strategy

In order to save ground resources, for TT&C stations, this paper plans to give priority to B and C stations with multi-beam equipment. For the payload-management station, Beijing station should be taken into consideration in priority as the main station. Figure 5 shows the relationship between the number of nodes and the number of satellites that can meet the requirements of TT&C and payload management at the same time.

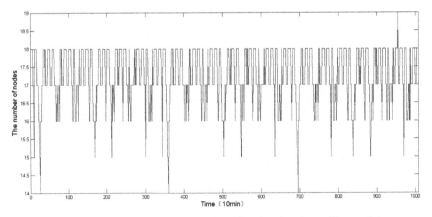

Fig. 5. The relation between the number of optional node satellites and time

From Fig. 5, it can be seen that the number of node satellites meeting the requirements of TT&C and payload management is 11–17 in the entire operation period of satellite constellation.

Selecting 11 Node Satellites

Assuming that the number of node satellites is 11, the non-node satellites cannot build links with the non-node satellites continuously when constructing timeslot table. In order to maximize the utilization of node satellites, non-node satellites adopt the scheduling method of build links alternately with node satellites and non-node satellites. There are 30 satellites in the whole GNSS constellation. According to the analysis of Sect. 3.2, during a time cycle, each non-node satellite has 10 discontinuous links with node satellites, at least 9 of which are different.

When the number of node satellites is 11, the ideal timeslot table can be obtained in the whole 7-day period, but there are some timeslots that cannot meet the time delay requirements, as shown in Fig. 6. The map shows the timeslot table of 8750 min– 8760 min. The X-axis represents 30 satellites and the Y-axis represents 20 timeslots. The red-filled satellites are node satellites and the rest are non-node satellites.

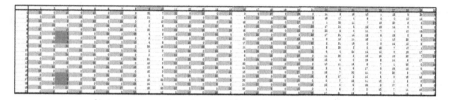

Fig. 6. The timeslot table of 8750 min–8760 min

As can be seen from the graph, for most satellites, they can meet the delay requirement of less than 3 s, but the green area of the figure shows that the delay is more than 3 s, which fails to meet the delay requirements.

For the whole 7-day period, about 36% of the timeslots do not meet the delay requirements in 1088 timeslots.

Selecting 12 Node Satellites

Figure 6 shows that Beijing Station cannot establish satellite-to-ground links with 12 node satellites at the same time in some time cycles, so it needs to add a payload-management station to increase the number of satellite-to-ground links.

In this paper, two combination schemes of "Beijing Station+A Station" and "Beijing Station+C Station" are used to establish the simultaneous time relationship diagram of the number of optional node satellites under different combination, as shown in Fig. 7. As can be seen from the graph, the number of node satellites in the former is more than that in the latter. Therefore, in order to use ground resources as few as possible, the scheme of Beijing and C's main and standby payload-management stations collocation is preferred.

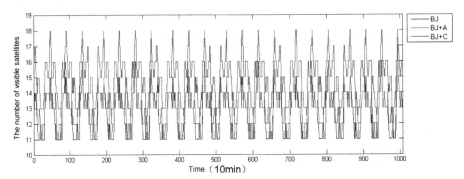

Fig. 7. Time relationship diagram of the number of optional node satellites under different combinations

Applying 12 node satellites, through traversing, the entire timeslot table across the cycle is obtained, and the timeslot table of A station in a time slice is randomly selected. As shown in Fig. 8, for non-node satellites, the time delay meets the requirement.

Fig. 8. Any part of the slot table when 12 nodes selected

For non-node satellites, the number of links built in every timeslot within a time cycle can be increased by building links with non-node satellites. The average number of visible and minimum number of non-node satellites is 10.55 and 8, which can fully meet the requirements of PDOP and the number of links.

Dynamic Min-Node-Satellites Selection Strategy

Through the analysis mentioned above, in order to optimize the utilization of ground resources more reasonably, we should adopt the optimal one which is called dynamic min-node-satellites selection strategy. That is, when using 11 node satellites can meet the requirements, then we adopt the above method, and 12 node satellites should be selected in other cases. In this way, it can meet the autonomous navigation requirements of inter-satellite links with as few ground resources as possible. According to the scheduling principle of preferential selection of multi-beam equipment, the corresponding visible matrix between ground resources and node satellites can be obtained. The results got by the above methods are simulated and analyzed, and the ground resources usage as shown in Table 2 can be obtained.

Table 2. Ground resources usage under dynamic min-node-satellites selection strategy

Mode	Station	Average number of tracking nodes	Maximum number of tracking nodes	Minimum number of tracking nodes	Longest tracking time (10 min)
TT&C	C	10.7480	12	8	1008
	B	0.8889	3	0	17
Payload management	Beijing+	10.7808	12	9	1008
	C	0.8562	3	0	16

5 Conclusion

In this paper, based on the networking requirements of the GNSS constellation and the current situation and development of the ground station network, according to the principle of using the multi-beam equipment preferentially, the extended services of the GNSS constellation are fully expending and the requirements of autonomous navigation and time delay are given priority. In the two cases, two node satellites selection strategies are proposed. By analyzing the ground resource usage of the two strategies, the optimal deployment plan of the ground resources is obtained. The results of this paper have certain reference significance for the subsequent route scheduling, and provides effective assistant decision for the optimization of the network scheme and satellites management next. With the continuous expansion of other services such as high-speed and low-speed data transmission of GNSS constellation and the continuous improvement of service quality, how to optimize link scheduling further based on the two strategies proposed in this paper is the key point to be followed.

References

1. Cao, Y., et al.: Initial analysis of the BDS satellite autonomous integrity monitoring capability. GPS Solut. **23**(2), 35 (2019)
2. Venkata Ratnam, D., et al.: Development of multivariate ionospheric TEC forecasting algorithm using linear time series model and ARMA over low-latitude GNSS station. Adv. Space Res. S0273117718302436 (2018)
3. Xiaoyong, S., Yue, M., Laiping, F., et al.: The preliminary result and analysis for BD orbit determination with inter-satellite link data. Acta Geodaetica Cartogr. Sin. **46**(5), 547–553 (2017)
4. Zixuan, L., Zhibo, Y., Kanglian, Z., et al.: Performance evaluation of routing algorithms in navigation constellation networks. J. Nanjing Univ. (Nat. Sci.) **3**, 530–536 (2018)
5. Jing, L., Ye, G.Q., Jiang, Z., et al.: A station routing algorithm satisfied ground station distribution constraint for satellite constellation network. In: Science and Information Conference(SAI) (2015)
6. Da-Cheng, L., Yan, L., Yan-Fei, L., et al.: Present status and development trends of inter-satellite link. Telecommun. Eng. **54**(7), 1016–1024 (2014)
7. Pan, J., et al.: Time synchronization of new-generation BDS satellites using inter-satellite link measurements [EB/OL]. Adv. Space Res. (2017). https://doi.org/10.1016/j.asr.2017.10.004

8. Razoumny, Y.N.: Fundamentals of the route theory for satellite constellation design for Earth discontinuous coverage. Part 3: low-cost earth observation with minimal satellite swath. Acta Astronaut. **129**, 447–458 (2016)

9. He, J., Jiang, Y., Zhang, G., et al.: Topology and route production scenario of Walker satellite constellation network with inter-satellite link. J. PLA Univ. Sci. Technol. (Nat. Sci. Edn.) **05** (2009)

10. Jing, L., Ye, G.Q., Jiang, Z., et al.: A routing algorithm satisfied ground station distribution constraint for satellite constellation network. In: Science and Information Conference (SAI) (2015)

11. Teunissen, P.J.G.: Integer least-squares theory for the GNSS COMPASS. J. Geodesy, July 2010

12. Huang, J., Su, Y., Liu, W., et al.: Optimization design of inter-satellite link (ISL) assignment parameters in GNSS based on genetic algorithm. Adv. Space Res. **60**(12), S0273117716307499 (2016)

13. Zhou, J.: An improved satellite link scheme with beam restriction for the navigation constellation. Sci. Sinica **41**(5), 575–580 (2011)

14. Sun, J., Hu, C., Li, Y., Lin, L.: The discussion on local optimization of navigation constellation based on STK/MATLAB. In: Sun, J., Yang, C., Guo, S. (eds.) CSNC 2018. LNEE, vol. 499, pp. 197–208. Springer, Singapore (2018). https://doi.org/10.1007/978-981-13-0029-5_18

15. Zhang, T.Y., Ye, G.Q., Li, J., et al.: FTS-based link assignment and routing in GNSS constellation network. In: China Satellite Navigation Conference (2017)

16. Teng, Y., Wang, J.: Some remarks on PDOP and TDOP for multi-GNSS constellations. J. Navig. **69**(01), 145–155 (2016)

17. Shi, L.Y., Xiang, W., Tang, X.M.: A station link assignment algorithm for GNSS with crosslink ranging. In: International Conference on Localization & GNSS (2011)

Research on Satellite Occurrence Probability in Earth Station's Visual Field for Mega-Constellation Systems

Ziqiao Lin, Wei Li, Jin Jin$^{(\boxtimes)}$, Jian Yan, and Linling Kuang

Department of Electronic Engineering, Tsinghua University, Beijing 100084, China
617339261@qq.com, jinjin_sat@tsinghua.edu.cn

Abstract. In this paper, we focus on the satellite occurrence probability in the earth station's visual field for mega-constellation systems. Based on the configuration of Walker constellation, we deduce the geometric relationship between the satellite phase interval and the line-of-sight angle of earth station at different elevation angles, and then analyze the law between constellation scale and line-of-sight angle of adjacent satellites. According to the given line-of-sight angle, the visual field of earth station is divided into several regions, and the satellite occurrence probability in each region can be derived. The analysis indicates that, with the expansion of the constellation scale, satellite occurrence probability in each region increases and approaches to 1. Based on the above analysis, we propose an access and handover scheme suitable for mega-constellation systems. It is assumed that the earth station is equipped with a wide shaped beam, which is pointing to a certain fixed region. There is always a visible satellite in the beam, through which the earth station connects to the constellation system and it can avoid the calculation burden of extrapolating a large number of satellite orbits. Simulation results demonstrate that the variations of space-earth link distance, radial velocity and elevation angle are extremely slight. Therefore, our proposed scheme can be well applied in the mega-constellation with frequent handovers, which can reduce the overhead of time and frequency calibration.

Keywords: Mega-constellation · Occurrence probability · Access and handover scheme · Interference avoidance

1 Introduction

During the operation of the constellation system, the earth station may be in the coverage of multiple satellites at the same time. Access and handover scheme directly affect the performance of the constellation system [1, 2]. Traditional schemes include distance-prioritized, coverage time prioritized [3], load-balancing [4], average weighted [5], etc. The earth station has to build ephemeris and extrapolate the positions of all satellites

This work is supported by the National Nature Science Foundation of China (Grant No. 91738101) and Shanghai Municipal Science and Technology Major Project (Grant No. 2018SHZDZX04) and BNRist (Grant No. 20031887521).

© Springer Nature Singapore Pte Ltd. 2020
Q. Yu (Ed.): SINC 2019, CCIS 1169, pp. 207–220, 2020.
https://doi.org/10.1007/978-981-15-3442-3_17

at every moment in the implementation of the above scheme, which is suitable for the situation where the overlap of the satellite is rare (2-4 overlaps) and the number of satellites is small. At present, the growing number of mega-constellation plans are proposed: OneWeb plans to launch 1980 satellites [6] and Boeing has announced the deployment of 2946 satellites in v-band and c-band [7]. Meanwhile, Samsung plans to launch 4600 satellites in the future [8] and Space X will deploy 11943 low-Earth orbit (LEO) satellites [9]. These mega-constellations have far more satellites than traditional constellations, and the earth stations can be covered by dozens or even hundreds of satellites. If the earth station adopts the traditional schemes and extrapolates the satellite orbits, it will lead to a large amount of computation and high computational complexity. In addition, ultra-high coverage overlaps will lead to frequent handover, resulting in a lot of overhead such as communication protocol, timing and frequency adjustment, which will seriously reduce the working efficiency of the satellite system.

With the advent of the era of 5G and Internet of things (IoT), the scale of Cell of Origin is shrinking. The access of massive data is bound to bring frequent handovers, and the ground communication network often improves the handover efficiency by simplifying communication protocol. Pure ALOHA protocol is a simple communication protocol [10] that sends data whenever the user has it, but it is prone to data conflicts, so the throughput is low. Slotted ALOHA is an improvement of pure ALOHA protocol [11], which only sends data at the beginning of the time slot and improves system throughput. Azari A *et al.* proposed Grant Free, a communication protocol for short packets. By removing synchronization/reservation requirements, multiple copies of packet are sent at the transmitter side and more complex signal processing at the receiver side [12], so as to achieve the purpose of reducing the cost of communication. In the traditional constellation system, due to the large spatio-temporal variation of the space-earth link, the communication link can only be established by calibrating time and frequency for each handover, so the handover procedure cannot be simplified by referring to the ground protocol.

As the scale of the constellation increases, the number of satellites in the visible field of the earth stations increases from several to dozens or even hundreds, the satellite density enhanced, and the probability of satellites occurrence in various points in the visible field increases correspondingly. Based on the mega-constellation scene, this paper establishes the geometric model of the relative position among earth station, satellite and earth center, and deduces the relationship among the elevation of earth station, the line-of-sight angle of adjacent satellites and the satellite phase interval. The results show that the probability of satellite occurrence with low elevation is higher than that with high elevation. With the expansion of constellation scale, the probability with various points in the visible field of earth station will approach to 1.

Based on the above derivation, we propose an access and handover scheme for mega-constellation. It is assumed that the shaped-beam of the earth station is wide, and the visible field of the earth station is divided into regions, so that the beam points to a certain fixed region. When the scale of the constellation is large enough, there is always a visible satellite in the beam, through which the earth station is connected to the constellation system. In Walker constellation, we simulate the occurrence probability of the divided regions in the visible field and verify the correctness of the model. The simulation results about radial velocity, elevation and the distance between the satellite and the earth station preliminarily show the feasibility of the scheme proposed in this paper.

Mega-constellations also need to consider the co-frequency problem. Currently, the most common methods to avoid interference are adjusting transmit power and setting the isolation angle. The latter means that the angle between different beam directions is greater than a certain value. In practice, the method of setting the isolation angle is more suitable [13–16]. According to the scheme proposed in this paper, the beam pointing of the earth station remains unchanged, while the direction of the space-earth link remains fixed. By coordinating the beam direction with the adjacent earth stations, the co-frequency interference can be avoided.

2 Scenario and System Model

2.1 Configuration and Principle of Walker Constellation

In the 1970s, British Walker and American Ballard proposed the concept of the Walker constellation. It is adopted by many constellations because of its good global coverage and latitude coverage [17, 18].

Walker constellation is described by three parameters: $T/P/F$, T represents the total number of constellation satellites, P represents the number of orbital planes, and F is the phase factor, which determines the initial phase difference $\Delta\omega_f = 2\pi \cdot F/T$ of adjacent satellites in adjacent orbits. Generally, combined with the orbital altitude h and inclination i, the positions of all satellites in Walker constellation can be completely determined [18].

With the expansion of constellation scale, the earth stations covered overlaps by satellites increases. The beam angle of earth station of LEO satellite constellation is large, typically 15–20°. The relationship among the elevation of the earth station, the line-of-sight angle of the adjacent satellites and the satellite phase interval in Walker constellation is derived below.

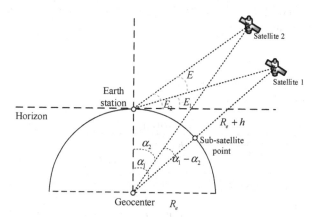

Fig. 1. Relationship among the elevation, the satellite phase interval and the line-of-sight angle

As is shown in Fig. 1, E_1 and E_2 are the elevation of satellite 1 and satellite 2 respectively, and α_1 and α_2 are the geocentric angle, and h is the orbital altitude, and R_e is the earth radius.

The relationship between the elevation and the geocentric angle is known from [18]:

$$\alpha_1 = \arccos[\frac{R_e}{h + R_e} \cos E_1] - E_1 \tag{1}$$

$$\alpha_2 = \arccos[\frac{R_e}{h + R_e} \cos E_2] - E_2 \tag{2}$$

According to the geometric relations in Fig. 1:

$$E_2 - E_1 = E \tag{3}$$

$$\alpha_1 - \alpha_2 = \varphi \tag{4}$$

where E is the line-of-sight angle of the earth station, $\varphi = 2\pi/(T/P)$ is the satellite phase interval on the same orbital plane. The following formulas can be obtained from (1), (2), (3) and (4):

$$\varphi = \arccos[\frac{R_e}{h + R_e} \cos E_1]$$
$$- \arccos[\frac{R_e}{h + R_e} \cos(E + E_1)] + E \tag{5}$$

$$E = \arctan[\frac{(h + R_e)\cos(\alpha_1 - \varphi) - R_e}{(h + R_e)\sin(\alpha_1 - \varphi)}]$$
$$- \arctan[\frac{(h + R_e)\cos\alpha_1 - R_e}{(h + R_e)\sin\alpha_1}] \tag{6}$$

(5) and (6) are the relationship between E_1, φ and E. When E and E_1 are given, the satellite phase interval can be determined. When the beam angle of the earth station β satisfies the condition:

$$\beta > E \tag{7}$$

there must be a visible satellite in beam range.

The partial derivative of Eq. (5) to E_1:

$$\frac{\partial \varphi}{\partial E_1} = \frac{R_e \cdot \sin E_1}{(R_e + h) \cdot \sqrt{1 - R_e^2 \cdot \cos^2 E_1/(R_e + h)^2}}$$
$$- \frac{R_e \cdot \sin(E_1 + E)}{(R_e + h) \cdot \sqrt{1 - R_e^2 \cdot \cos^2(E_1 + E)/(R_e + h)^2}} \tag{8}$$

When $E_1 \in [\theta, \frac{\pi}{2} - \frac{E}{2}]$, $\frac{\partial \varphi}{\partial E_1} < 0$, in which θ is the minimum elevation of the earth station. With the increase of E_1, φ decreases (the satellite number in same orbit increases), and the magnitude of the decrease slows down gradually.

The results indicate that when the satellite number is constant and elevation is low, the probability of satellite occurring in the beam is greater. With the increase of elevation, the probability of satellite occurrence decreases, and the decrease rate slows down. As the constellation scale continues to increase, the probability of satellite occurrence will continue to increase and eventually approach to 1.

2.2   Access and Handover Scheme

Based on the above results, this paper proposes an access and handover scheme for mega-constellation system. We consider a Walker constellation, the visible field of the earth station is divided into several regions. According to the given center interval, we calculate the probability of satellite occurrence in each divided region. As the scale of the constellation expands, the probability of satellite occurrence in various regions will increase continuously and approach to 1 gradually. It is assumed that the earth station's shaped-beam is wide, when the conditions of communication are met, the beam can be directed to a region with a high probability of occurrence and keep pointing to the fixed direction. Since there is always at least one satellite in the beam, the earth station can access the constellation system through that satellite.

Inter-satellite handover is required when the access satellite is about to leave the pointing region. We assume that there are inter-satellite links between adjacent satellites. The data of the satellite, which is about to be moved out of the region, can be transmitted to other satellites in the same region via the inter-satellite link. The beam of the earth station is still pointing to this region (or fine-tuned) and accesses the new satellite to continue communication.

When multiple earth stations are densely distributed, they can be coordinated with each other or be uniformly scheduled by the satellite control center, so that the beams of the earth stations always point to different regions, which have similar probabilities, to achieve spatial separation and avoid co-frequency interference (Fig. 2).

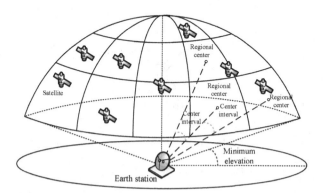

Fig. 2. The division regions of the earth station's visible field

3 Occurrence Probability Calculation Method

Based on the Walker constellation, this section discusses the calculation method of satellite occurrence probability. In this section, we already know: the period of satellite and the specific position at each moment, the location information of earth stations and the minimum elevation, the regional center interval.

Under the above assumptions, we calculate the probability of satellite occurrence in each partition region. It should be noted that each earth station has the same method, we only provide the method to calculate the probability in the division regions by calculating a certain earth station $u_i (i = 1, 2, \cdots, U$, where U is the total number of the earth station).

3.1 Calculate the Positions of the Visible Satellites

The satellite position at the initial moment, total satellite number S and satellite period T can be obtained through the built-in ephemeris of the earth station. We assume that s is time step and take M periods to calculate, $k = M \cdot T / s$ is the number of moments of M periods, and extrapolate the position of the satellite at k moments by using the initial orbital conditions.

For each moment, we judge whether satellite $s_j (j = 1, 2, \cdots, S)$ is visible to the u_i within the minimum elevation θ. The formula is as follows:

$$\arccos \left(\frac{\vec{P}_{u_i} \cdot \vec{P}_{u_i s_j}}{\left| \vec{P}_{u_i} \right| \left| \vec{P}_{u_i s_j} \right|} \right) < \theta \tag{9}$$

where $\vec{P}_{u_i}$ is the vector pointing from the origin to u_i and $\vec{P}_{u_i s_j}$ is the vector of u_i pointing to satellite s_j (Fig. 3).

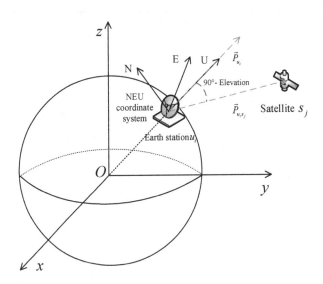

Fig. 3. ENU coordinate system and elevation

When the satellite is in the visible field of the earth station, we calculate azimuth and elevation of visible satellite s_j relative to u_i. We transform $\vec{P}_{u_i s_j}$ to a new vector $\vec{P}'_{u_i s_j}$ in the East-North-Up (ENU) coordinate system with the u_i as the origin. The elevation can be obtained by calculating the complementary angle of the angle between $\vec{P}'_{u_i s_j}$ and the U-axis; the azimuth can be obtained by calculating the angle between the projection of $\vec{P}'_{u_i s_j}$ on the EON plane and the E-axis.

Through iterative computation, we get the elevation $E_{s,n \times k}$ and azimuth $A_{s,n \times k}$ of the visual satellite at k moments, where n is the maximum number of visible satellites.

3.2 Division Region of Visual FieldB

We assume that γ is the central interval of each region. The visible field is divided by the direction of the U-axis, i.e., the elevation of the earth station is 90°. The regional center is projected in the two-dimensional rectangular coordinate system plane (as shown in Fig. 4), where the X-axis direction is east, the Y-axis direction is north, and the origin is directly above the earth station.

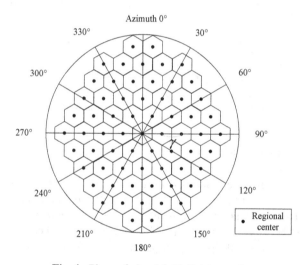

Fig. 4. Plane of visual field division region

We calculate the angle of the regional center, origin and X-axis, and obtain the azimuth $A_{N_1 \times 1}$ of the regional center, in which N_1 is the total number of regions. We get the elevation $E_{N_1 \times 1}$ by calculating the distance between the regional center and the origin.

Taking the value greater than the minimum elevation θ in $E_{N_1 \times 1}$, and take the corresponding value in $A_{N_1 \times 1}$ to get the azimuth $A_{N \times 1}$ and the elevation $E_{N \times 1}$ in the center of regions, in which N is the number of regions in the visible field.

3.3 Probability of Satellite Occurrence in the Divided Region

Through the above method, we get the elevation $E_{s,n \times k}$ and the azimuth $A_{s,n \times k}$ of the visual satellite relative to the earth station u_i at k moments, and the $A_{N \times 1}$ and $E_{N \times 1}$ of the divided region center. In which the azimuth ranges from $0°$ to $360°$ and the elevation from θ to $90°$.

We get the distance between two points according to the cosine formula:

$$d = \left(\left(\frac{\pi}{2} - e_i \right)^2 + \left(\frac{\pi}{2} - e_{s,j,t} \right)^2 \right.$$
$$\left. -2 \cdot \left(\frac{\pi}{2} - e_i \right) \cdot \left(\frac{\pi}{2} - e_{s,j,t} \right) \cdot \cos(a_i - a_{s,j,t}) \right)^{\frac{1}{2}} \tag{10}$$

in which $a_i \in A$, $e_i \in E(i = 1, 2, \ldots, N)$, $a_{s,j,t} \in A_s$, $e_{s,j,t} \in E_s$, where $a_{s,j,t}$ represents the azimuth of the j th visible satellite at time t, e_{sjt} represents the elevation of the j th visible satellite at time t. We judge which region center is the closest to the j th satellite at the current moment, then the j th satellite is in this region.

Recording whether satellites appear in each region at k moments, and obtain a matrix $L_{N \times k}$. Further, we can get the average probability of satellite occurrence in each region as:

$$P_{N \times 1} = \frac{\sum\limits_{j=1}^{k} l_{i,j}}{k} \tag{11}$$

Where $l_{i,j} (i = 1, 2, \ldots, N, j = 1, 2, \ldots, k)$ is the value in $L_{N \times k}$.

4 Simulation and Analyses

In this section, we will give the simulation results and analyze the law of occurrence probability and the feasibility of the access and handover scheme. The simulation parameters are as follows. The constellation scale is increased by the number of orbital satellites as twice the number of orbital planes, and 10 satellites are added each time; the visible field is divided according to the triangle arrangement (Table 1).

Table 1. Simulation parameter

Parameter	Value
Orbital altitude h	770 km
Orbital inclination i	88°
Number of satellites S	200, 450,..., 11250
Number of orbital planes P	10, 15,..., 75
Phase factor F	1
Simulation period M	5

(continued)

Table 1. (*continued*)

Parameter	Value
Comparison period number m	5
Earth station longitude Lon	116.388°
Earth station latitude Lat	39.9289°
Regional center interval γ	20°
Minimum elevation θ	15°
Time step s	2 s
Satellite period T	6014 s
Number of regions N	53

The simulation evaluation index is variance, the calculation formula of average variance is:

$$\sigma_{ave} = \frac{\sum_{i=1}^{m} \left(P_{ave,N\times1} - P_{i,N\times1} \right)^2}{m} \tag{12}$$

where $P_{ave,N\times1}$ is the average probability of satellite occurrence calculated within the M period, and each row represents the average probability of satellite occurrence in each region, $P_{i,N\times1}$ is the average occurrence probability of the i th computing period, m is the number of comparison period (Fig. 5).

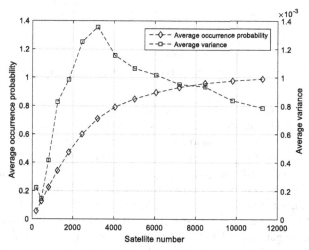

Fig. 5. Average variance and average satellite occurrence probability increase with constellation number

The simulation results show that, as the number of satellites gradually increases, the average probability of satellites occurrence in the divided region tends to 1, i.e., there are satellites in every region at any moment. When the number of satellites is small, the average probability of occurrence is slightly higher than 0, which indicates that the probability in each divided region is very low. As the number of satellites increases, the average probability of occurrence increases gradually, and the fluctuation of satellite occurrence probability increases in each region. When the average probability of occurrence is close to 1, the variance decreases gradually, indicating that the satellites in every divided region tend to be constant. For earth station, satellites will continue to appear in the divided region, the beam of the earth station can point to a certain region continuously, or fine-tune the direction in this region, and access satellites in this region.

It can be seen from Fig. 6 that the maximum probability of occurrence approaches to 1 when the total number of satellites reaches 3,200 satellites, and the minimum probability of occurrence is about 0.28. The relative relationship between the probability of occurrence at this moment and the location of the regional centers are shown in Fig. 7:

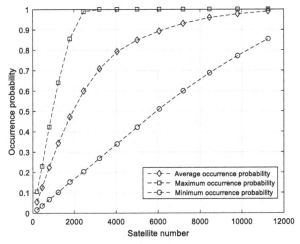

Fig. 6. Average/maximum/minimum occurrence probability as the number of constellations increases

As shown in Fig. 7, the position of the circle represents the regional centers, the polar angle 0° is north, and the length of the polar axis represents the elevation. The coordinate center is directly above the earth station, i.e., the elevation is 90°. From Fig. 7, we can see that the minimum probability of satellites occurrence is above the earth station. As the divided regions expand to the periphery, the probability of satellite occurrence gradually increases, which is the conclusion derived from Sect. 2. The probability of occurrence in the north is higher than that in the south because the Walker constellation is more densely distributed at the poles and sparsely distributed toward the equator. Through Eq. (5), in this configuration, when the occurrence probability of the minimum elevation is 1, the number of satellites per orbit is 48, which corresponds to the simulation results. The line-of-sight angle can be obtained by Eq. (6). It is known from (7) that, at this elevation, if the beam angle is larger than the line-of-sight angle, at least one satellite can be seen within the range of the beam angle.

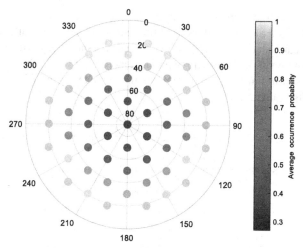

Fig. 7. Relationship between the average occurrence probability of 3200 satellites and the regional centers

During the satellite handover process, the earth station antenna needs to rotate direction to re-align a new satellite, hence the earth station will be unable to communicate for a period of time. We assume that under the shortest access scheme, the antenna needs to rotate for an average of 100 s to re-align a satellite, while 150 s under the longest access time scheme. Figure 8 shows the ratio of the communication duration to the total duration. It can be seen from Fig. 8 that other access schemes may not be able to communicate for a relatively fixed period of time. However, the access time of the scheme proposed in this paper will gradually increase with the expansion of constellation scale. In other words, in mega-constellation scenarios, our scheme can ensure the continuity of communication without communication interruption caused by handover.

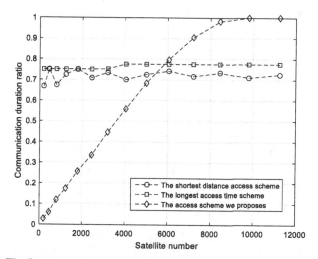

Fig. 8. Communication duration ratio of different access schemes

Under three constellation scale of 200, 3200 and 11250, we calculate the radial velocity, distance and elevation of the satellite relative to a certain earth station during the period from earth station access satellite to handover satellite. We take the maximum value and minimum value of these three parameters respectively, subtract the minimum value from the maximum value, and calculate the average value of 30 times to get the following results (Table 2).

Table 2. Radial velocity, distance, elevation of different constellations

Satellite number	Parameter	Value
200	Radial velocity	2.32 km/s
	Distance	444.37 km
	Elevation	28.07°
3200	Radial velocity	1.48 km/s
	Distance	41.85 km
	Elevation	13.17°
11250	Radial velocity	0.86 km/s
	Distance	12.0 km
	Elevation	7.44°

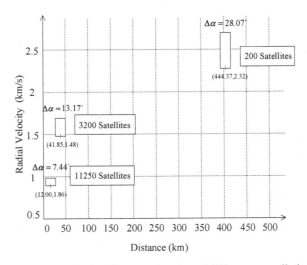

Fig. 9. Radial velocity, distance, elevation of different constellations

It can be seen from Fig. 9 that as the number of constellation satellites increases, the three parameters variation of the satellite decreases. The elevation, space-earth link distance, radial velocity correspond to channel quality, link loss and Doppler frequency offset respectively. These communication indicators hardly change when the constellation scale is large enough. At this point, if the access and handover scheme proposed in this paper is adopted, and the handover procedure is simplified by referring to the ground protocol, the overhead such as time and frequency calibration during the handover procedure can be omitted.

5 Conclusions

In this paper, we study the probability of satellite occurrence in the visual field of mega-constellation system. The analysis indicates that with the expansion of constellation scale, the probability of satellite occurrence in each region of the visible field will increase continuously and approach to 1. Based on the above conclusions, we propose an access and handover scheme suitable for mega-constellation systems. The simulation result shows the feasibility of this scheme and indicates that it is suitable for mega-constellations with frequent handover.

This scheme also considers the problem of co- frequency interference. When multiple earth stations are densely distributed, they can be coordinated with each other, so that the beams of earth stations always point to different regions, which have similar probabilities, to achieve spatial separation and avoid co-frequency interference.

References

1. Wang, C.Q.: Research on Handover Scheme of LEO Satellite Communication System. Beijing University of Posts and Telecommunications (2018)
2. Wang, Y.H.: Research on Capacity Analysis and Access Control Strategy of Non-geostationary Orbit Satellite Communication System. University of Electronic Science and Technology of China (2018)
3. Ling, X., Hu, J.H., Wu, S.Q.: Access schemes for LEO satellite mobile communication systems. Acta Electronica Sinica **28**(7), 70–73 (2000)
4. Evangelos, P., Stylianos, K., Gerasimos, D., et al.: Satellite handover techniques for LEO networks. Int. J. Satell. Commun. Netw. **22**(2), 231–245 (2010)
5. Hang, F., Xu, H., Zhou, H., et al.: Qos average weighted scheme for LEO satellite communications. J. Electron. Inf. Technol. **10**, 2411–2414 (2008)
6. Sohail, D.: UK OneWeb To Launch 1,980 Satellites To Orbit To Provide Global Internet Access President [EB/OL]. https://www.urdupoint.com/en/technology/uk-oneweb-to-launch-1980-satellites-to-orbit-538352.html
7. Irwin, D.: Boeing Developing V-band Satellite Constellation [EB/OL]. https://www.radioworld.com/trends-1/boeing-developing-vband-satellite-constellation
8. Borghino, D.: Samsung's giant satellite network could enable high-speed internet access across the globe [EB/OL]. https://newatlas.com/samsung-satellite-network-internet-access/38971/
9. Albulet, M.: SPACE X Technical Information to Supplement Schedle. Space Exploration Technologies Corp., 1 March 2017
10. Li, J.D., Sheng, M., Li, H.Y.: Communication Network Foundation, vol. 5, 2nd edn. Higher Education Press, Beijing (2011)
11. Roberts, L.G.: ALOHA packet system with and without slots and capture. ACM SIGCOMM Comput. Commun. Rev. **5**(2), 28–42 (1975)

12. Amin, A., Petar, P., Guowang, M., et al.: Grant-Free Radio Access for Short-Packet Communications over 5G Networks (2017)
13. Jin, J., Li, Y.Q., Zhang, C., et al.: Occurrence probability of co-frequency interference and system availability of non-geostationary satellite system in global dynamic scene. J. Tsinghua Univ. (Sci. Technol.) **58**(9), 833–840 (2018)
14. Zhang, C., Jiang, C.X., Jin, J., et al.: Spectrum sensing and recognition in satellite systems. IEEE Trans. Veh. Technol. **68**, 2502–2516 (2019)
15. Zhang, C., Jin, J., Kuang, L.L., et al.: Blind spot of spectrum awareness techniques in non-geostationary satellite systems. IEEE Trans. Aerosp. Electron. Syst. **54**, 3150–3159 (2018)
16. Zhang, C., Jin, J., Zhang, H., et al.: Spectral coexistence between LEO and GEO satellites by optimizing direction normal of phased array antennas. China Commun. **15**(6), 18–27 (2018)
17. Song, Z.M., Dai, G.M., Wang, M.C., et al.: Theoretical analysis of walker constellation coverage to area target. Comput. Eng. Des. **10**, 3639–3644 (2014)
18. Zhu, L.D., Wu, T.Y., Zhuo, Y.N.: Introduction to Satellite Communication, vol. 3, 4th edn. Publishing House of Electronics Industry, Beijing (2015)

Coalition Formation Games for Multi-satellite Distributed Cooperative Sensing

Yunfeng Wang[1,2]([✉]), Xiaojin Ding[2], and Gengxin Zhang[1,2]

[1] National Engineering Research Center,
Nanjing University of Posts and Telecommunications, Nanjing 210003, China
2018010212@njupt.edu.cn
[2] Telecommunication and Network National Engineering Research Center,
Nanjing University of Posts and Telecommunications, Nanjing 210003, China

Abstract. In this paper, we propose a distributed cooperative sensing strategy for multi-satellites based on coalition formation games. To reduce the complexity, we introduce adjunct utility function to achieve steady-state coalition. The simulation results show that the merger-division-adjust (MDA) algorithm proposed in this paper can reduce the probability of missed detection compared with the non-cooperative algorithm. Compared with merger and division (MD) algorithm, MDA algorithm can reduce the probability of false alarm, indicating that the proposed strategy has better spectrum sensing performance and more suitable for satellites.

Keywords: LEO network · Multi-satellite coordination · Spectrum sensing · Coalition formation games

1 Introduction

Satellite communications [1] is playing more and more significant role in the future information transmission, which is an important supplement to ground 5G network. Among different types of satellite network, Low Earth Orbit (LEO) network provides global communication with short end-to-end delays, small transmission loss, low power dissipation. Moreover, the constellation composed of several LEO satellites can achieve seamless coverage of the whole world (including two poles). Thus, LEO satellite constellation networks are expect to play a significant role in the future satellite communications. Some companies have proposed large-scale constellation systems composed of hundreds or thousands of LEO satellites, such as SpaceX and OneWeb [2]. However, as continuously increasing demand for broadcast, multimedia and interactive services, the useable satellite spectrum is becoming scarce. It is extremely urgent to improve the spectrum utilization in satellite communications.

This work is supported in part by the National Science Foundation of China under Grant 91738201, and Grant 61801445.

© Springer Nature Singapore Pte Ltd. 2020
Q. Yu (Ed.): SINC 2019, CCIS 1169, pp. 221–229, 2020.
https://doi.org/10.1007/978-981-15-3442-3_18

For efficiently exploiting these spectrum holes, cognitive radio has been proposed as a promising technology to enhance the efficiency of spectrum and make full use of available spectrum resources [3–5]. Cooperative spectrum sensing, as one of the key techniques of CR, has been paid more attention, especially in the field of satellite communication. A novel protocol focused on the problem of hidden incumbent during network entry and handover for satellite based on cognitive radio is proposed in [6]. Spectrum sensing algorithms in cognitive satellite networks are summarized, especially the implementation of spectrum sensing technique are introduced in [7]. The author in [8] studies an optimal transmission control method in cognitive wireless network for satellite networks. And it can be concluded that the spectrum efficiency can been obviously improved with the CR techniques in satellite communication [9].

Most of existed works are based on non-fading channel [10]. However, when the ground frequency equipment is located in the urban areas or the elevation is small, the shadow fading is more serious, and the sensing performance based on this channel can not meet the actual needs. In [11], the author analyses the performance of spectrum sensing system in Rician fading channel and Nakagami channel. The LOO fading model is studied in [12], which can be very close to the satellite-to-ground propagation environment, but it hard to be calculated. In [13], author propose a new shadowed Rice model for land mobile satellite channels, which provides a similar fit to the experimental data as the well-accepted Loos model but with significantly less computational burden.

The rest of this paper is organized as follows. Section 2 presents the system model for Shadowed-Rician fading channel. In Sect. 3, the proposed coalitional game approach is described. Simulation results are given in Sect. 4. Finally, we conclude the paper in Sect. 5.

2   System Model

In this paper, we consider a cognitive radio network for satellite communications with one primary user and N satellites as secondary users, as shown in Fig. 1 The satellites adopt energy detectors, which is one of the main practical signal detectors in cognitive radio networks.

2.1   Local Sensing

Each satellite performs spectrum sensing independently, the average probability of false alarm probability (P_f^i) is still the same as that under the non-fading channel due to P_f^i is independent of signal-to-noise ratio (SNR), given by:

$$P_f^i = P_f = Pr(\lambda|H_0) = \frac{\Gamma(u, \frac{\lambda}{2})}{\Gamma(u)} \tag{1}$$

Where $\Gamma(u, \frac{\lambda}{2})$ refers to incomplete gamma function.

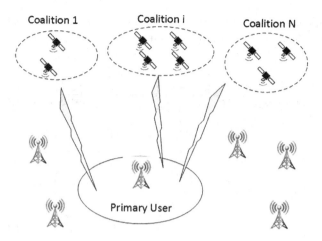

Fig. 1. Illustrative of distributed cooperative sensing model for satellite communications

And the average probability of detection (P_d^i) of satellite i with the energy detector, can be obtained by averaging over fading statistics,

$$P_d^i = \int_{\gamma_i} \overline{P_d^i} \, f(\gamma_i) \, d\gamma_i = \int_{\gamma_i} Q_u(\sqrt{2\gamma_i}, \sqrt{\lambda_i}) \, f(\gamma_i) \, d\gamma_i \qquad (2)$$

Where λ_i is detection threshold for the satellite i, $Q_u(u,v)$ represents the Marcum-Q function.

In this paper, satellite channel is as Shadowed-Rician fading channel. The probability distribution function (PDF) of power channel gain $|h_i(t)|^2$ under the aforementioned fading model is given by :

$$f_{|h_i|^2}(x) = \alpha_i e^{-\beta_i x} {}_1F_1(m_i; 1; \delta_i x), x \geq 0 \qquad (3)$$

where $\alpha_i = (2b_i m/2b_i m + \Omega)^m/2b_i$, $\delta_i = \Omega/((2b_i m + \Omega)2b_i)$, $\beta_i = 1/2b_i$, the parameter Ω represents the average power of the LOS component, $2b_i$ is the average power of the scatter component, and m is the Nakagami parameter ranging from 0 to ∞. For $m = 0$, the envelope of $|h_i(t)|^2$ obeys the Rayleigh distribution, which is associated to urban region with complete obstruction of the line-of-sight (LOS). While for $m = \infty$, it follows the Rician distribution, which is associated to open region with no obstruction of the LOS. Furthermore, ${}_1F_1(m_i; 1; \delta_i x)$ is the confluent hypergeometric function.

Given γ is the SNR and $\gamma_s = |h_i|^2\gamma$, the PDF of γ_i can be shown as:

$$f_{\gamma_i}(x) = \frac{1}{\gamma} \times f_{|h_i|^2}(\frac{x}{\gamma}) = \frac{\alpha_i}{\gamma} exp(-\frac{\beta_i}{\gamma}x) {}_1F_1(m_i; 1; \frac{\delta_i x}{\gamma}), x \geq 0 \qquad (4)$$

Hence, the closed-form formula for P_d under Shadowed-Rician fading channel can be obtained (after some manipulation) by substituting $f(\gamma_i)$ into Eq. 2, yielding:

$$
P_d^i = \frac{1}{2\gamma b_i} \left(\frac{2b_i m}{2b_i m + \Omega}\right)^m \times \sum_{n=0}^{\infty} \left\{ \frac{(m)_n \Omega^n}{(1)_n n! (2\gamma b_i)^n (2b_i m + \Omega)^n} \right.
$$
$$
\times \left[\Gamma(n+1)(2b_i \gamma)^{n+1} - exp(-\frac{\lambda}{2}) \right.
$$
$$
\times \sum_{j=u}^{\infty} \left[\frac{\lambda^j \Gamma(n+1) 2^{n+1-j} (b_i \gamma)^{n+1-j}}{(2b_i \gamma + 1)^{n+1} \Gamma(j+1)} \right.
$$
$$
\left. \left. \left. \times \, _1F_1\left(n+1; j+1; \frac{2\lambda b_i \gamma}{4b_i \gamma + 2}\right) \right] \right] \right\}
$$

(5)

Where $(m)_n = \Gamma(m+n)/\Gamma(m)$ is an incremental factorial whose length is n. Moreover, an important metric is the probability of missing the detection of the PU, defined as:

$$
P_m^i = 1 - P_d^i
$$

(6)

2.2 Data Fusing

We assume that satellites have inter-satellite links and can exchange information with each other. Within each coalition $S \subseteq N$, a satellite designated as coalition leader collects which can fuse all available decision information to infer the absence or presence of the PU. For combining the sensing bits and making the final detection decision, the coalition leader uses the decision fusion OR-rule. Hence, the miss and false alarm probabilities for a coalition S, respectively, given by:

$$
Q_m^i = \prod_{i \in \Psi} [P_m^i (1 - P_e^{i,k}) + (1 - P_m^i) P_e^{i,k}]
$$

(7)

$$
Q_f^i = 1 - \prod_{i \in \Psi} [(1 - P_f^i)(1 - P_e^{i,k}) + P_f^i P_e^{i,k}]
$$

(8)

where $P_e^{i,k}$ is the probability of error due to fading on the reporting channel between the satellite i of coalition S and the coalition leader k. Without loss of generality, we assumed binary phase-shift-keying modulation, and the average probability of reporting error between satellites shown as,

$$
P_e^{i,k} = \frac{1}{2} erfc\left(\sqrt{\frac{E_b}{N_0}}\right)
$$

(9)

Where $P_e^{i,k}$ is a complementary error function, $\frac{E_b}{N_0}$ is expressed as the ratio of signal energy per bit to noise power, which can be obtained from the carrier-to-noise ratio, is given by:

$$P_e^{i,k} = \frac{1}{2}erfc\left(\sqrt{\frac{B_n}{R_b}\frac{C}{N}}\right) = \frac{1}{2}erfc\left(\sqrt{\frac{B_n}{R_b}\frac{P_tG_tG_r}{kT_sB_nL_fL_a}}\right) \tag{10}$$

Where R_b represent baud rate of signal, P_t represents the transmit power, G_t and G_r represents the gain of antenna, L_f represents the free space propagation loss, L_a represents other losses, T_s represents the equivalent noise temperature of the receiver, B_n represents the transponder bandwidth, and k is Boltzmann constant.

3   Collaborative Spectrum Sensing as Coalitional Game

3.1   Problem Formulation

The main purpose of satellite spectrum sensing is to find spectrum holes, as well as obtain frequency information in time domain, frequency domain, energy and so on. Therefore, the objective of optimization is to obtain the best sensing performance, which means to minimize the total error rate. Strictly speaking, the total rate of a satellite define as $P_1P_m^i + (1 - P_1)P_f^i$, where P_1 represents the probability that PU is occupying the spectrum. For conciseness, we assume $P_1 = 1/2$, and thus, the total error rate is given by $(P_m^i + P_f^i)/2$. However, it should be clear from Eqs. 7 and 8, as the number of cooperative satellites increases, the probability of false alarms increases while the probability of missed detection decreases. This tradeoff impact on the total error rate, and thus, our optimal objective is to characterize the network structure that will form when the satellites collaborate while accounting for this tradeoff. Moreover, we consider the average sensing performance of all satellites in the network while maintaining the probability of false alarm below a certain threshold. By omitting the factor, we have the following optimization problem:

$$\min_{T \in B} \frac{\Sigma_{\Psi \in T}(Q_m^i + Q_f^i)_{\Psi}}{N}, \qquad s.t. Q_f^i \leq \alpha \tag{11}$$

3.2   Coalitional Utility Function

Apparently, the number of possible coalition structures (partitions) is exponentially grows with N, problem 11 that the optimal coalition structure is intractable in the general case. Therefore, we consider suboptimal solution with distributed algorithm for satellites. To simplify the problem, we refer to coalitional game theory for spectrum sensing among the satellites, which makes the satellites can make autonomous decisions to join a coalition, and form the suitable coalitional structure.

By defining a coalition as a set of satellites sensing the same PU, we model the coalitional game as a nontransferable utility (NTU), and the function of the value, which quantifies the worth of a coalition in a game, is define as follows:

$$U(S^i) = 1 - (Q_m^i + Q_f^i) \tag{12}$$

The main purpose of coalitional game is to reduce the total error rate as much as possible through multi-satellites collaborative sensing. Hence, by proposed this algorithm, the satellite will continue to play with other satellites to improve the perceived performance, and continuously join or leave the alliance. This will increase the complexity of on-board processing and increase the burden of satellite processing. However, considering the satellite as a resource and energy-constrained platform, the on-board processing capability cannot be too complicated. As the fact of that, we set a desire detection value ε, once a coalition achieves the probability of miss ε, it will have no incentive to pursue the coalition formation. Hence, this can reduce the complexity of the algorithm and the overhead of the satellite while satisfying the spectrum sensing requirements. Therefore, we introduction adjunct utility function $V(S^i)$ as follows:

$$V(S^i) = \begin{cases} 1, & if \quad Q_m^i < \varepsilon \\ 0, & otherwise \end{cases} \tag{13}$$

3.3   Coalitional Game Formulation

We propose an algorithm that consists of three stages: (1) initial state, (2) coalition formation, (3) coalition sensing. In the initial state, all satellites are partitioned $\Upsilon = \{\Upsilon_1, \Upsilon_2, \cdots, \Upsilon_k\}$. To ensure perceived fairness, the number of satellites in each initial alliance is as equal as possible. Then every satellite performs local spectrum sensing, and exchanges information to partners through the inter-satellite link. The satellite with the lowest probability of missed detection is the leader of the coalition and calculates the utility value of the current alliance. Followed an adjust operation is applied to the coalition in the network. In this way, the initial coalitions $\{\Upsilon_1, \Upsilon_2, \cdots, \Upsilon_k\}$ be divided into two collections, one is the set W_0 with steady-state coalitions, and the other set L_0 is the remaining satellites.

In the coalition formation state, the satellite in L_0 selects the nearest coalition to perform the merge and split operation. If a satellite can improve its individual utility without decreasing the others utilities, it will decide to merge. Nonetheless to see, whenever a coalition forms by merge (or by split), it is subject to an adjust operation. Repeat this process and iterate until the coalition utility no longer grows. Consequently, similar to the initial coalition Υ, the set L_0 also be divided into two collections, one is the set W_f with steady-state coalitions, and the other set L_f is the remaining satellites.

From this definition, it follows that, when a coalition satisfies the probability of miss ε, it is no longer to pursue the coalition formation, which means a significant reduction in computation overhead.

In coalition sensing state, satellites that belong to the same coalition will report their local sensing bits to the leader of the coalitions through the inter-satellite link. The leader of the coalition will fuse the received information according to OR-rule and make the final decision on the presence or absence of the PU.

4 Simulation Results and Analysis

Our research is the multi-satellites cooperative sensing, which focuses on the performance analysis of the coalitional formation games algorithm. Therefore, for the simulations, we consider a network that 9 satellites are distributed in the area of 33 km, and the PU is 10^6 m to the satellite. The transmit power is set to 100 mw, noise power $\sigma_s^2 = -90$ dBm, the fading channel parameters are set as $(b, m, \Omega) = (0.126, 10.1, 0.835)$. In addition, the threshold of the adjunct utility function is set $\varepsilon = 0.05$. We simulate the multi-satellite cooperative sensing based on two algorithms. The first only contains the merge and split operation (MD), and the other is proposed in this paper, compared with the MD algorithm, add the adjustment operation.

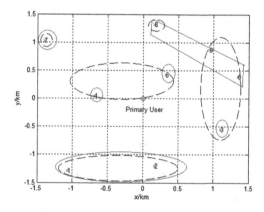

Fig. 2. Snapshot of the coalition structure for N = 9 satellites.

In Fig. 2, we show a snapshot of the coalitional structure resulting from both MD (in dashed line) and MDA (solid line) algorithms for N = 9 satellites, $P_f = 0.0128$, and a miss detection probability $Q_m^i = 0.05\%$ for MDA. The simulation shows that both strategies can form collaborative coalitions. According to the MDA, the satellites {3, 4, 5} will not participate in the coalition formation process. However, for the satellite {8}, in MD strategy, there is no coalition with other satellites, while in the MDA algorithm, the satellite form a coalition with the satellites {6, 9}. Finally, we note that for both MD and MDA, coalition {1, 2} forms, and this coalition is unable to find any suitable partners that can help it improve its utility further in MD or meet the target detection threshold in MDA. Therefore, we can conclude that compared with the MD strategy, the

MDA strategy proposed can adjust the coalition formation process based on the perceived task, which is more flexible and efficient.

Figures 3 and 4 show, respectively, the average probabilities of miss and the average false alarm probabilities achieved per satellite for different network sizes. In Fig. 3, we show that, both algorithms outperform the local spectrum sensing, and their miss probability reduce with the network size. The gap between MD and MDA is not obviously, indicating the performance of the two algorithms is similar. In Fig. 4, the false alarm probability in non-cooperative case does not change with the network size increasing, and outperforms the both two algorithm. Moreover, the false alarm probability in the MDA algorithm outperforms the MD algorithm in all cases, and the gap between them increase with the network size. When the network is sparse, both algorithm has similar performance. While, when the network is dense, the false alarm probability is reduced from 0.155 to 0.105, which is 67%. The reason is that according MDA, when

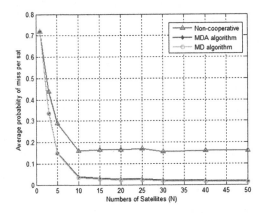

Fig. 3. Average probability of miss versus number of satellites (N).

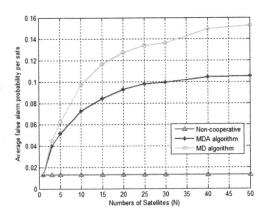

Fig. 4. Average false alarm probability versus number of satellites (N).

a coalition satisfies the probability of miss, the satellite is no longer to pursue the coalition formation, thus reducing the inter-satellite communication, and the false alarm probability is also reduced.

5   Conclusion

This paper has proposed a distributed cooperative sensing algorithm for multi-satellite based on coalition formation games. We introduce adjunct utility function to reduce the complexity of algorithm. The simulation results show that the MDA algorithm proposed in this paper can reduce the probability of missed detection compared with the non-cooperative algorithm. Compared with MD algorithm, MDA algorithm can reduce the probability of false alarm, indicating that the proposed strategy has better spectrum sensing performance and more suitable for satellite environment.

References

1. Qu, Z., Zhang, G., Xie, J.: LEO satellite constellation for Internet of Things. IEEE Access **85**, 18391–18401 (2017)
2. Snchez, A.H., Soares, T., Wolahan, A.: Reliability aspects of mega-constellation satellites and their impact on the space debris environment. In: Reliability and Maintainability Symposium, pp. 1–5. IEEE Press, Piscataway, NJ (2017)
3. Darcy, L.B., William, A.W., Aaron, W.: NASA's advanced tracking and data relay satellite system for the years 2000 and beyond. Proc. IEEE **78**(7), 1141–1151 (1990)
4. Lee, Y.: Modified myopic policy with collision avoidance for opportunistic spectrum access. Electron. Lett. **46**(10), 871872 (2010)
5. Tsiropoulos, G.I., Dobre, O.A., Ahmed, M.H., et al.: Radio resource allocation techniques for efficient spectrum access in cognitive radio networks. IEEE Commun. Surv. Tutor. **18**(1), 824–847 (2016)
6. Gozupek, D., Bayhan, S., Alagoz, F.: A novel handover protocol to prevent hidden node problem in satellite assisted cognitive radio networks. In: 2008 ISWPC 2008. 3rd International Symposium on Wireless Pervasive Computing (2008)
7. Sharma, S.K., Chatzinotas, S., Ottersten, B.: Cognitive radio techniques for satellite communication systems. In: 2013 IEEE 78th Vehicular Technology Conference (VTC Fall). IEEE (2013)
8. Uchida, N., Sato, G., Takahata, K., Shibata, Y.: Optimal route selection method with satellite system for cognitive wireless network in disaster information network. In: 2011 IEEE International Conference on Advanced Information Networking and Applications (AINA) (2011)
9. Hoyhtya, M., Kyrolainen, J., Hulkkonen, A., Ylitalo, J., Roivainen, A.: Application of cognitive radio techniques to satellite communication (2012)
10. Proakis, J.G.: Digital Communications. McGraw-Hill, New York (2009)
11. Digham, F.F., Alouini, M.S., Simon, M.K.: On the energy detection of unknown signals over fading channels, pp. 21–24. IEEE (2007)
12. Loo, C.: A statistical model for a land mobile satellite link. IEEE Trans. Veh. Technol. **34**(3), 122–127 (1985)
13. Abdi, A., Lau, W.C., Alouini, M.S., et al.: A new simple model for land mobile satellite chan-nels: first- and second-order statistics[J]. IEEE Trans. Wirel. Commun. **2**(3), 519–528 (2003)

Research on Satellite Communication System for Interference Avoidance

Feng Liu[1], Man Su[2], Jiuchao Li[1(✉)], Yaqiu Li[1], and Mingzhang Chen[1]

[1] Institute of Telecommunication Satellite, China Academy of Space Technology, Beijing 100094, China
lijiuchao@yeah.net

[2] Beijing Institute of Tracking and Telecommunication Technology, Beijing 100094, China

Abstract. The wireless transmissions of the satellite communications system are vulnerable to the interferences. In realistic systems, the satellite communications system faces three kinds of interference scenarios. Built upon the analysis on the possibly occurred three interference scenarios, we propose interference avoidance oriented cognitive satellite communications architecture such that the reliable communications can be guaranteed under the spectrum resource contentions and complex electromagnetic environments. Following the developed architecture, we further analyze the corresponding interference avoidance approaches. The work conducted in this paper could support the construction and development of future cognitive satellite communications systems.

Keywords: Cognitive radio · Cognitive satellite communication · Interference avoidance

1 Introduction

Satellite communication system an irreplaceable role in the communication supporting system due to its good characteristics such as wide coverage, good transmission quality, rapid deployment, convenient networking. Meanwhile, the investment for communication system is almost independent of communication distance, and the target destinations of satellite communication are almost not limited by the conditions of geographical environment. With the explosive growth in data communication services and the unceasing application of various wireless communication modes, the terrestrial and space-based spectrum resources are becoming increasingly crowded and competing, which aggravates the deterioration of the electromagnetic environment of the space-based information transmission. To make full use of the limited spectrum resources and ensure the reliable communication of the satellite communication system in the complex electromagnetic environment of spectrum competition and congestion, lots of domestic and overseas scholars are integrating the concept of cognitive radio into the satellite communication system. Cognitive radio (CR) technology, which was first proposed by Dr. Joseph Mitola based on software defined radio (SDR) in 1999, is a new technology to

F. Liu and M. Su—These authors made equal contributions to this work.

© Springer Nature Singapore Pte Ltd. 2020
Q. Yu (Ed.): SINC 2019, CCIS 1169, pp. 230–241, 2020.
https://doi.org/10.1007/978-981-15-3442-3_19

improve the utilization efficiency of wireless spectrum resources. It allows unauthorized users to dynamically access idle spectrum so that the spectrum utilization can be effectively improved. There are four major parts that constitute CR technology, i.e., spectrum sensing, spectrum analysis, spectrum decision and spectrum handoff [1, 2]. Based on the CR technology oriented for satellite communication, we can design the so called satellite communication system based on cognitive radio, which is briefly called cognitive satellite communication system. Many domestic and foreign scholars as well as industrial institutions are trying to find and define the model of cognitive satellite communication system. In 2010, S. Kandeepan et al. [3] introduced the concept that satellite terrestrial cognitive radio can dynamically access the spectrum of terrestrial system in hybrid satellite terrestrial system, and proposed the innovative application of cognitive radio satellite, which is called cognitive satellite terrestrial radio. The project team CoRaSat (cognitive radio for satellite communications) supported by ETSI (European Telecommunications Standards Institute) has been launched to study the satellite communication technology equipped with cognitive functions. In the project of CoRaSat, the satellite communication with cognitive function are allowed to use the idle licensed spectrum of authorized satellite for communication on the premise of not interfering with the normal work of the authorized satellite users, so as to make full use of the limited spectrum resources [4]. The establishment of the CoRaSat project team marks that the design of cognitive satellite communication system architecture is becoming the focus of researches all over the world, which means that the standardization related to the top-level design of the cognitive satellite communication system has started. In 2014, ETSI developed the standard on the applications of cognitive radio in satellite communication, known as Cognitive Radio Techniques for Satellite Communications Operating in Ka Band (ETSI TR 103 263 v1.1.1), and three most advantageous cognitive satellite communication scenarios were recognized based on the practice of the CoRaSat and extensive market analysis. Although CR technology in satellite communication system has not been widely studied, some research institutions and universities have already carried out exploratory researches. Shanghai Institute of Microsystem and Information Technology (SIMIT) has studied the feasibility and implementation conditions of CR technology in the field of satellite communication. Researchers from SIMIT are devoted to the research of low orbit satellite communication system, and have conducted thorough analysis on the state of the art and application prospect of CR key technologies in satellite communication network according to the characteristics and demands of satellite system. China Academy of Space Technology (Xi'an) has carried out research and simulation on the application prospect of spectrum sensing technology of CR for satellite systems [5]. In this research, researchers established the cognitive radio access model of satellite and optimally design the spectrum sharing and power allocation technology. The research shows that CR technology is conducive to alleviating the current under-utilization of spectrum resources and weak anti-interference ability of satellite communication system, as well as improving the throughput of satellite communication network [6]. The National Natural Science Foundation of China (NSFC) and the National High Technology Research and Development Program(863 Program) of China attach great importance to the research of CR, and the NSFC Information Sciences Department established a number of key projects and project groups in the field of CR, mainly including cognitive wireless mesh network routing research, cognitive-based wireless resource management and utilization, as well as the theory and key technologies of multi-user cooperative diversity in CR system. In recent years, scholars at home and abroad have done a lot of research work in cognitive satellite communication system architecture, spectrum sensing algorithm and cognitive satellite communication system network, providing theoretical support for the construction of cognitive satellite communication system [7–10].

Above all, it can been seen that the concept of cognitive satellite communication system mainly stays in the researches of academia and the exploration of industrial departments, most of which are demand analysis, scenario description and tentative researches on the top-level design of the system in the field of communication satellite. Through the summary and reasoning of foreign strategic objectives and technical research abroad, the United States has at least carried out relevant concept researches, and relevant information shows that the United States has already enable the terrestrial user terminals of MUOS communication satellite system to have certain spectrum sensing ability, and take some measures to eliminate interference on the basis of sensing. Currently, the construction goal and research direction of cognitive satellite communication system has turned into improving the anti-interference ability of the system. This capability is realized through the joint spectrum sensing and dynamic spectrum resource allocation. When designing the system, we need to focus on the overall planning, architecture design and compatible coexistence between the system itself and other communication systems as well as equipment using frequency. However, the interference caused by other equipment using frequency outside the whole system architecture is very complicated, and it is very difficult to accurately predict and identify the interference signals. Whether the accurate prediction, identification, evaluation and decision-making for interference signal can be carried out, as well as whether the further corresponding dynamic spectrum access management can be flexibly and rapidly realized are the keys to the success of avoiding interference for the system. Each subsystem of the integrated satellite and terrestrial cooperative communication system carries out spectrum sensing within its own frequency band, then confirms the available frequency set and conducts frequency negotiation, and finally selects a pair of uplink and downlink frequencies to establish the connection for communication. In the process of communication, the system can even continue spectrum sensing and perform real-time channel handoff to avoid interference. With CR technology, even when 99% of the available frequencies of the whole cognitive satellite communication system are blocked, the only channels can still be utilized to establish the corresponding communication links. Thus, the capability of CR technology is obviously better when compared with other interference avoidance technologies. In this paper, we analyze three kinds of interference scenarios, propose a cognitive satellite communication system architecture based on CR technology, and investigate the working pattern for the interference avoidance in cognitive satellite communication systems.

2   Analysis of Interference Scenarios in Cognitive Satellite Communication System

Due to the wide coverage of satellites, the coverage includes not only a large number of satellite system user terminals and equipment, but also different types of terrestrial wireless communication systems using licensed frequency bands, such as broadcast television systems, cellular mobile communication systems, and etc. At the same time, satellite communication systems may also face interference from other systems in the operating band. According to different interference scenarios, the following three interference scenarios are analyzed.

2.1   Scenario 1: Satellite Uplink Interference

The operating band of the satellite uplink collides with the frequency band of the terrestrial system, and the terrestrial system transmits signals to interfere with the satellite uplink. To this end, it is necessary to perform spectrum sensing on the satellite uplink to find idle spectrum resources that are not occupied by the terrestrial system at the satellite receiving end for communication. In the interference scenario shown by Fig. 1, a spectrum sensing processor is configured in a satellite (or a station) to detect an interference signal of the uplink, locate the interfered frequency band, and discover the idle frequency band. When the user terminal initiates an access request through the uplink control channel, the satellite (or the station) completes the allocation of the uplink traffic channel according to the spectrum sensing result and other constraints on the resource scheduling. Then, the satellite (or the station) transmits the allocation result to the user terminal through the downlink control channel to complete the spectrum allocation. At the same time, after the establishment of the traffic channel, the satellite resource management module can dynamically adjust the current uplink channels to implement dynamic management of the spectrum resource according to the variation of the interference in uplink sensing results.

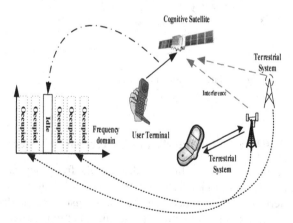

Fig. 1. Satellite uplink interference scenario

2.2   Scenario 2: Satellite Downlink Interference

The operating band of the satellite downlink collides with the frequency band of the terrestrial system, and the terrestrial system transmits signals to interfere with the satellite downlink. The user terminal needs to perform spectrum sensing on the downlink to find idle spectrum resources for communication. These idle spectrum resources are not occupied by the terrestrial system around the geographical location of the user terminal. In the interference scenario shown by Fig. 2, a spectrum sensing processor is configured in the user terminal to detect the interference signal in the downlink, locate the interfered frequency band, and discover the idle frequency band. When the user terminal initiates an access request through the uplink control channel, the spectrum sensing result of the user terminal needs to be sent to the satellite (or the station). Then the satellite (or the station) performs the allocation of the downlink traffic channel according to the spectrum sensing result and the other constraints on the resource scheduling, and sends the allocation result to the user terminal through the downlink control channel to complete the spectrum allocation. At the same time, after the establishment of the traffic channel, the user terminal can also transmit the variation of interference in the downlink sensing result to the satellite (or the station) in time, and the satellite (or the station) dynamically adjusts the current downlink channels according to the interference variation to implement dynamic management of the spectrum resource.

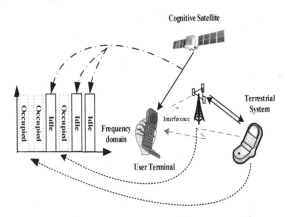

Fig. 2. Satellite downlink interference scenario

2.3   Scenario 3: Satellite Uplink Interferes with Terrestrial System

The operating band of the satellite uplink collides with the frequency band of the terrestrial system, and the satellite uplink interferes with the signal transmitted by the terrestrial system. The user terminal needs to perform spectrum sensing on the downlink to find idle spectrum resources for communication. These idle spectrum resources are not occupied by the terrestrial system around the geographical location of the user terminal. In the interference scenario shown by Fig. 3, the spectrum sensing processor is configured in the user terminal to detect the uplink interference signal around its location, locate the interfered frequency band, and discover the idle frequency band. When the user terminal initiates an access request through the uplink control channel, it needs to send its spectrum sensing result to the satellite (or station). Then the satellite (or the station) completes the allocation of the uplink traffic channel according to the spectrum sensing result and other constraints on the resource scheduling, and sends the allocation result to the user terminal through the downlink control channel to complete the spectrum allocation. At the same time, after the establishment of the traffic channel, the user terminal can also transmit the variation of interference in the uplink sensing result to the satellite (or the station) in time, and the satellite (or the station) dynamically adjusts the current uplink channels according to the interference variation to realize dynamic management of the spectrum resource.

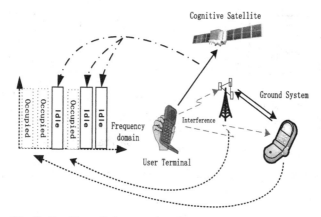

Fig. 3. Satellite uplinks cause interference to the ground system

3   Cognitive Satellite Communication System Architecture for Interference Avoidance

The cognitive satellite communication system architecture for interference avoidance is shown in Fig. 4, which includes satellites, user terminals and stations.

(1)   The main functions of the satellite are as follows:

Satellite uplink spectrum sensing, satellite uplink and downlink spectrum resource allocation, dynamic spectrum geographic information database update, transmitting uplink and downlink spectrum allocation results to user terminals, forwarding satellite uplink spectrum to the station, and forwarding user terminal spectrum sensing results to the station, and transmitting the satellite spectrum sensing results to the station.

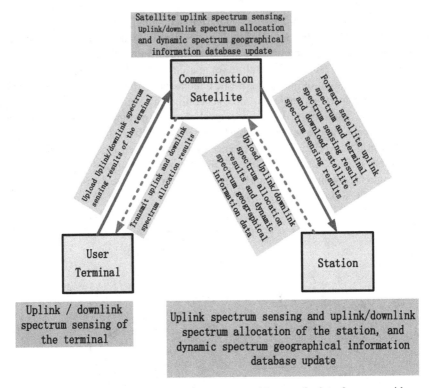

Fig. 4. Cognitive satellite communication system architecture for interference avoidance

(2) The main functions of the station are as follows:

Uplink spectrum sensing of the station, uplink and downlink spectrum resource scheduling of the station, dynamic spectrum geographical information database update, uploading uplink and downlink spectrum allocation results to the satellite, and transmitting dynamic spectrum geographical information to the satellite.

(3) The main functions of the user terminal are as follows:

Uploading the results of uplink and downlink spectrum sensing of the user terminal to satellite.

4   The Working Pattern of the Cognitive Satellite Communication System for Interference Avoidance

As shown in Fig. 5, the working pattern of the cognitive satellite communication system for interference avoidance can be divided into four phases: sensing phase, sensing result transmission phase, resource allocation (channel allocation) phase, and allocation result transmission phase.

Sensing phase: The satellite performs spectrum sensing on the uplink channels of the repeater, and user terminals perform spectrum sensing on the uplink and downlink channels at its geographical location.

Sensing result transmission phase: The spectrum sensing result of the satellite is sent to the station through the feeder link with an interaction mechanism, and the sensing result of user terminal is fed back to the station by letting the satellite forward the feedback with the interaction mechanism.

Resource allocation (channel allocation) phase: According to the sensing result, the station performs the allocation of the overall resources in combination with the data of the dynamic spectrum geographical information database, and figures out the availability of

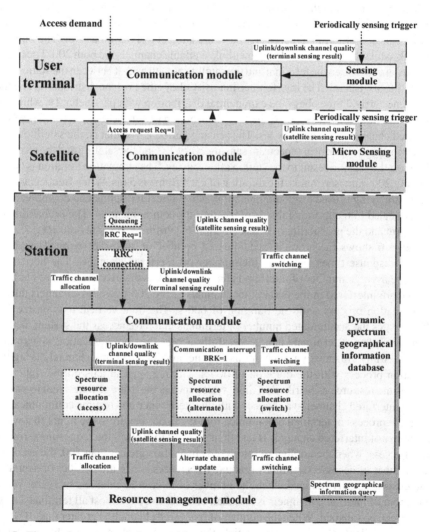

Fig. 5. The work pattern of the cognitive satellite communication system for interference avoidance

each channel. The traffic channel is allocated in real time when the user terminal initiates an access request, or when the user terminal initiates a traffic channel switching request in face of interferences on the traffic channel.

Allocation result transmission phase: the station sends the resource allocation result to the user terminal through the feeder link under an interaction mechanism with the satellite forwarding the resource allocation result to the user terminal.

5   Simulation Results and Analysis

To verify the performance of the above system architecture, a cognitive satellite communication system simulation platform is built for simulation evaluation. The simulation parameters are as follows: there are 200 terminals within one beam coverage, and the transmission of terminal access requests are subject to Poisson arrival. The number of uplink available channels and the downlink available channels are both 200. The transmission delay between the terminal and the satellite as well as that between the station and the satellite are 125 ms. The interference intensity obeys the Poisson distribution, and the interference arrival time obeys the exponential distribution with parameter 2 s, while the interference duration obeys the exponential distribution within the range of [6, 12] and the interfered channel updates every 8 s. The sensing period is 7 s, that is, the satellite communication system performs spectrum sensing every 7 s, and the satellite and the terminal will report the sensing result to the station after sensing. The simulation duration is 600 s, including 200 frames/s, that is, the duration of each frame is 5 ms. When frequency division multiplexing is used for communication, one channel for uplink and downlink need to be allocated to the terminal that is requesting for communication. The probability of false alarm and the probability of miss detection spectrum sensing are both 1%.

Figure 6 shows the how the accumulated number of accessed terminals and the resource response times vary with the number of interfered channels. The horizontal axis is the average number of interfered channels which is the average number of channels that are interfered in the 200 uplink channels or the 200 downlink channels during the simulation process with the duration of 600 s. The blue vertical axis on the left side indicates the accumulated number of terminals that have access the system, that is, the total number of terminals that successfully access the communication system and are successfully allocated with the uplink channel and the downlink channel during the simulation process with the duration of 600 s. The red vertical axis on the right side indicates the resource redistribution time, which means the average time spent to switch from an interfered channel to a channel without interference and resume communication during the process of terminal communication. It can be seen from the figure that when the number of interfered channels is less than or equal to 100, all the requesting terminal can get access; when the number of interfered channels is greater than 100, the accumulated number of accessed terminals gradually decreases, since the number of available channels is gradually reduced when the number of interfered channels increase. When the number of interfered channels is less than or equal to 100, almost all terminal access requests can be satisfied. When the number of interfered channels is greater than 100, the number of available channels cannot satisfy the user's access demand and the number of accessed users gradually decreases as the interference increases. To avoid interference, the satellite is supposed to forward the switching request to the station, and then the station will allocate available resource to the satellite and the terminal, which cause a delay of 500 ms, so the response time of resource redistribution must be greater than 0.5 s. With increasing interference, the number of available channels is reduced, and it takes much longer to find another available channel for the interfered channel, so the response time of resource redistribution increases as the number of interfered channels increases.

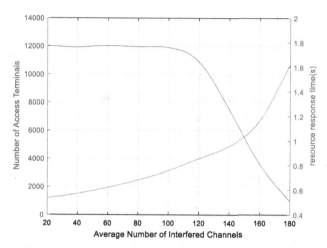

Fig. 6. Number of access terminals and the resource redistribution response times with average number of interfered channels (Color figure online)

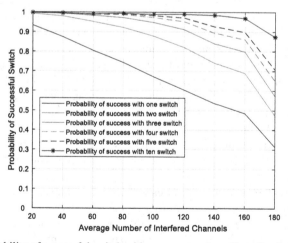

Fig. 7. The probability of successful switch with average number of interfered channels when the number of switch is 1,2,3,4,5,10.

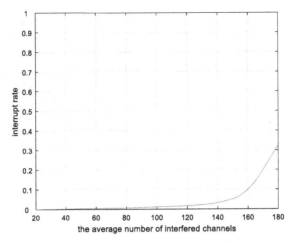

Fig. 8. The interrupt rate with average number of interfered channels

Figure 7 shows how the probability of successful switch varies with average number of interfered channels. The horizontal axis is the average number of interfered channels, while the vertical axis is the probability that the communication can be successfully switched to the channel without interference after the communication is interfered, wherein the probability of successful switch is the rate of total number of successful switch to total number of switching requests. The curves in the figure indicate the probability of successful switch when the number of switching is 1, 2, 3, 4, 5, as well as the final probability of successful switch, respectively. According to the protocol design, in one communication process, the switching request will be sent when the channel is interfered and up to ten alternate channels will be searched to ensure the communication to be carried out as far as possible. The communication will be interrupted if ten searching times run out so that the final probability of successful switch is the same as the probability of successful switch when the number of switching is 10. It can be seen from the Fig. 7 that as the number of interfered channels increases, i.e., the quality of channel degrades, the probability of successful switch gradually decreases, but the final probability of successful switch can be maintained at a high level, and it can still achieve 87% in the case of heavy interference. In addition, when the average number of interfered channels is less than 160, the final probability of successful switch remains above 98%, while the probability of success with one switch is significantly reduced, i.e., as the number of interfered channels increases, it will take more switching times to successfully switch to the available channels, which corresponds to a longer channel response time in Fig. 6.

Figure 8 shows the interrupt rate varies with average number of interfered channels. The vertical axis indicates the probability that the terminal will be interrupted due to failure of switching to the available channel during the communication. The interrupt rate is the rate of total number of interruptions to the total number of accesses. It can be seen from the Fig. 8 that as the number of interfered channels increases, the interrupt rate gradually increases. The interrupt rate increases sharply after the number of interfered channels is greater than 140. This is because there are too many interfered channels and a little available channels. When the communicating user is disturbed, it is difficult to find an available channel for it, so the interrupt rate dramatically increase.

6 Conclusion

This paper analyzed the cognitive satellite communication network system, especially for three kinds of interference scenarios of cognitive satellite communication system. Based on three scenarios of cognitive satellite communication system, we designed a cognitive satellite communication system workflow. In this workflow, we designed the working pattern of cognitive satellite communication system for interference avoidance, and built a cognitive satellite communication system level simulation platform to simulate the performance of the proposed system workflow. The development of current cognitive satellite communication systems is ambitious. In all possible frequency ranges, using coordinated and unified protocol mechanisms, real-time database management as well as efficient and accurate spectrum sensing technology to achieve dynamic spectrum access management will greatly enhance the system's capability of interference avoidance. With the rapid development of satellite communication technology, there are various categories of interferences to the satellite communication systems. The research on interference avoidance technology and anti-interference technology for satellite communication system will be an important guarantee for reliable communication of satellite communication system in the complex electromagnetic environment under spectrum competition and congestion.

References

1. Mitola, J.I., Maguire, G.Q.: Cognitive radio: making software radios more personal. IEEE Pers. Commun. **6**(4), 13–18 (1999)
2. Brown, T.X.: An analysis of unlicensed device operation in licensed broadcast service bands. In: IEEE International Symposium on New Frontiers in Dynamic Spectrum Access Networks, pp. 11–29 (2005)
3. Kandeepan, S., Nardis, L., De Benedetto M.G., et al.: Cognitive satellite terrestrial radios. In: IEEE Global Communications Conference, pp. 1–6 (2010)
4. Liolis, K., Schlueter, G., Krause, J., et al.: Cognitive radio scenarios for satellite communications: the CoRaSat approach. In: Future Network and Mobile Summit, pp. 1–10 (2013)
5. Chen, P., Qiu, L.D., Wang, Y.: Joint optimization algorithm of detection threshold and power allocation for satellite underlay cognitive radio. J. Xi'an Jiao tong Univ. **47**(6), 31–36 (2013). (In Chinese)
6. Chen, P., Qiu, L.D., Wang, Y.: Uplink power allocation of satellite underlay cognitive radio. Appl. Electron. Tech. **38**(12), 109–113 (2012). (In Chinese)
7. Jia, M., Liu, X., Yin, Z., et al.: Joint cooperative spectrum sensing and spectrum opportunity for satellite cluster communication networks. Ad Hoc Netw. **58**(1), 231–238 (2017)
8. Abdel-Rahman, M.J., Krunz, M., Erwin, R.: Exploiting cognitive radios for reliable satellite communications. Int. J. Satell. Commun. Netw. **33**(3), 197–216 (2015)
9. Kolawole, O.Y., Vuppala, S., Sellathurai, M., et al.: On the performance of cognitive satellite-terrestrial networks. IEEE Trans. Cogn. Commun Netw. **3**(4), 668–683 (2017)
10. Chae, S.H.: Cooperative communication for cognitive satellite networks. IEEE Trans. Commun. **66**(11), 5140–5154 (2018)

Constant Envelope Rate Compatible Modulation

Feng Feng, Yuqiu Zhou, Yu Zhao, Fang Lu[✉], and Yan Dong

School of Electronic Information and Communications,
Huazhong University of Science and Technology, Wuhan 430074, China
lufang@mail.hust.edu.cn

Abstract. The conventional rate compatible modulation (RCM) uses the high-order quadrature amplitude modulation (QAM) signal for rateless transmitting. The large size complex constellation with near-Gaussian probability mass function (PMF) produces high peak-to-average power ratio (PAPR). This paper presents a new method aimed at solving the PAPR problem associated with RCM. We transform the RCM signal to a constant envelope signal through concatenating RCM with continuous phase modulation (CPM), which decreases the PAPR to 0 dB. At the receiver, the serial iterative demodulating and decoding procedure is designed to improve the performance of the system. In the presence of nonlinear power amplifier, we simulate the bit error rate and spectral efficiency of the RCM-CPM and the RCM-QAM in additive Gaussian noise channels. The simulation results demonstrate that RCM-CPM outperforms RCM-QAM with input back-off, especially the performance advantage is about 3 dB at high signal-to-noise ratios (SNRs).

Keywords: Continuous phase modulation · Rate compatible modulation · Peak-to-average power ratio · High signal-to-noise ratios

1 Introduction

In wireless networking, rate adaptation is critical to the system performance by exploiting the dynamic bandwidth. Rate compatible modulation (RCM) [1, 2] is attractive for achieving seamless and blind rate adaptation under time varying channel. It does not require the knowledge of channel at the sender and minimizes the feedback requirement to one-bit acknowledgment. Furthermore, RCM achieves near-capacity performance in a wide range of channel signal-to-noise ratios (SNRs) and is insensitive to the channel estimation error. Owing to these advantages, RCM is a competitive technology that can be widely applied in rate adaptive communication systems.

The multi-level RCM coded symbols are generated via a sparse mapping matrix, which leads to the occurrence of RCM coded symbols are with near-Gaussian probability mass function (PMF). The conventional RCM uses the high-order quadrature amplitude modulation (RCM-QAM) signal for transmitting, every two consecutive coded symbols

---

This work is supported by the National Nature Science Foundation of China (91538203).

© Springer Nature Singapore Pte Ltd. 2020
Q. Yu (Ed.): SINC 2019, CCIS 1169, pp. 242–254, 2020.
https://doi.org/10.1007/978-981-15-3442-3_20

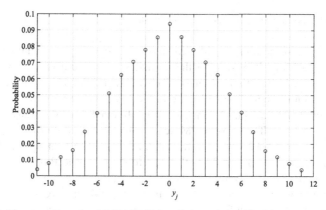

Fig. 1. The near-Gaussian PMF of RCM coded symbols, $\mathbf{W} = \{\pm 1, \pm 2, \pm 4, \pm 4\}$

create one complex-valued constellation symbol. Due to the large size complex constellation with unequal probability, the primary drawback of RCM-QAM is that the modulated waveform has large amplitude fluctuations and produces high peak-to-average power ratios (PAPRs). The high PAPR makes RCM-QAM sensitive to nonlinear power amplifier (PA). Without effective power back-off, the system suffers spectral expansion and performance degradation. Although increasing the input back-off (IBO) can alleviate these problems, but meanwhile, PA efficiency is reduced. When RCM is applied to a satellite communication system, due to the long communication distance, the system requires large input power and has high linearity requirement for nonlinear amplifiers. It is more urgent to solve the high PAPR problem of RCM.

At present, there is only one literature [3] on reducing the PAPR of RCM, which proposes a novel non-linear constellation mapping (NLCM) scheme. In NLCM, each RCM coded symbol is mapped to a complex-valued amplitude phase shift keying (APSK) constellation point. However, NLCM could only improve the spectral efficiency and reduce the PAPR for low-to-moderate SNRs. NLCM still has as high as 10 dB PAPR to obtain high spectral efficiency at high SNRs, therefore, the large IBO is needed.

In addition, orthogonal frequency-division multiplexing (OFDM) has high PAPR problem. Some techniques have been developed for OFDM to address the PAPR problem: distortion-less PAPR reduction schemes such as coding and tone reservation [4]; non-distortion-less PAPR reduction schemes such as clipping/filtering and peak windowing [5]; pre-distortion schemes [6] and constant envelope OFDM (CE-OFDM) [7]. Many of these methods can be applied to reduce PAPR of RCM with minor modification. In particular, CE-OFDM provides a post-processing technique for OFDM to achieve constant envelope modulation. The basic idea is to combine OFDM with phase modulation (PM). The advantage of CE-OFDM is that the transformed signal has the lowest achievable PAPR, 0 dB. Obviously, the significance of the 0 dB PAPR is that the signal can be amplified efficiently with nonlinear power amplifiers. However, due to the fact that the baseband signal of CE-OFDM has double sideband spectral, CE-OFDM's spectral efficiency is less than half of the original OFDM. In [8, 9], a dual-stream structure with single antenna is proposed to increase the spectral efficiency of CE-OFDM. This structure transmits dual-stream baseband at the same time so that the spectral efficiency is improved.

RCM can be also combined with PM to achieve envelope modulation, while the discontinuity phase of PM results in the reduction of spectral efficiency for integer output signal of RCM. Continuous phase modulations (CPM) [10] is another constant envelope modulation scheme. In CPM, the phase is a continuous function of time and each date symbol only affects the instantaneous frequency of the transmitted signal in one symbol interval. Therefore, CPM achieve higher spectral efficiency compared with PM. In addition, CPM is suitable for satellite communication with limited power. In this paper, we concatenate RCM with CPM to decrease the PAPR to 0 dB, which improves the nonlinear adaptability of RCM and promotes the application of RCM in satellite communication systems. As for combining CPM with RCM, the difficulty is how to design the way soft information is transmitted between component decoders. Hence, we adopt a serial iterative demodulating and decoding procedure with iteratively soft-input soft-output (SISO) modules [11, 12] at the receiver. The Balh-Cocke-Jelinek-Raviv (BCJR) algorithm is used for demodulating CPM and the belief propagation (BP) algorithm is used for decoding RCM, the soft information is exchanged between BCJR and BP to improve the performance.

The rest of this paper is organized as follows. Section 2 reviews the PAPR problem of RCM-QAM and presents the proposed RCM-CPM satellite system model. Section 3 presents the receiver structure of RCM-CPM, which adopts iterative demodulating and decoding. In the presence of nonlinear PA, the simulation results of the bit error rate (BER) and spectral efficiency performance are given in Sect. 4. Finally, the conclusions are drawn in Sect. 5.

2   General System Description

2.1   The PAPR Problem of RCM-QAM

Let $\mathbf{X} = \{x_1, x_2, \cdots, x_N\} \in \{0, 1\}^N$ represent an N-dimensional source bit vector, and $\mathbf{\Phi}$ represent an $M \times N$ low-density random mapping matrix, then an M-dimensional multi-level coded symbol vector $\mathbf{Y} = \{y_1, y_2, \cdots, y_M\}$ can be generated by weighted sum operations:

$$\mathbf{Y}^{\mathrm{T}} = \mathbf{\Phi}\mathbf{X}^{\mathrm{T}} \tag{1}$$

Each row of $\mathbf{\Phi}$ has only $n(n \ll N)$ entries that are nonzero and take values from a weight set $\mathbf{W} = \{w_1, w_2, \cdots, w_n\}$. In RCM, $\mathbf{W}$ is assumed to be symmetric, which means that for any $w_i \in \mathbf{W}$, we have $-w_i \in \mathbf{W}$.

In conventional RCM-QAM, every two consecutive coded symbols create one complex-valued constellation symbol s_m, as follows:

$$s_m = y_{2m-1} + \sqrt{-1}\, y_{2m}, m > 0 \tag{2}$$

which is sent through the channel.

Let K denote the number of all possible RCM coded symbols for given weight set $\mathbf{W}$, it is calculated by $K = \sum_{n'=1}^{n} |w_{n'}| + 1, w_{n'} \in \mathbf{W}$. For the considered weight set $\mathbf{W} = \{\pm 1, \pm 2, \pm 4, \pm 4\}$, which could achieve an overall high throughput for SNR range from 5 dB to 25 dB, $K = 23$ and RCM coded symbols take values from $-11-+11$. The PMF of RCM coded symbols is near-Gaussian, as shown in Fig. 1. Then, RCM-QAM achieves a 23×23 complex constellation with unequal probabilities.

Each source bit is either 0 or 1 with equal probability 0.5. Then, we can get the average signal power $E_{s0} = 1/2 \sum_{n'=1}^{n} w_{n'}^2$, and the maximum value of coded symbol is then $y_{max} = 1/2 \sum_{n'=1}^{n} |w_{n'}|$. The PAPR for RCM-QAM (without the raised cosine finite impulse response (FIR) filter), denoted by η_{so}, is given by

$$\eta_{s0} = 2y_{max}^2 / E_{s0} = \left(\sum_{n=1}^{n} |w_{n'}| \right)^2 / \sum_{n'=1}^{n} w_{n'}^2 \tag{3}$$

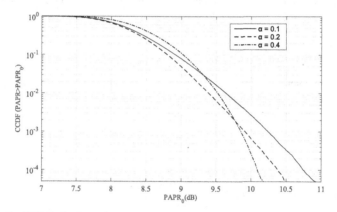

Fig. 2. CCDF of RCM with various roll-off factor α, $\mathbf{W} = \{\pm 1, \pm 2, \pm 4, \pm 4\}$

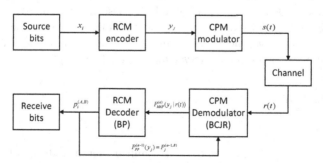

Fig. 3. The serially concatenated scheme of RCM-CPM

Due to the large size of the complex constellation and the near-Gaussian PMF of symbols, RCM-QAM has high PAPR. For the considered weight set $\mathbf{W} = \{\pm 1, \pm 2, \pm 4, \pm 4\}$, η_{so} is as high as 8.16 dB.

The QAM constellation symbols are shaped with raised-cosine FIR with roll-off factor α before transmitting. The relationship between α and complementary cumulative distribution function (CCDF) of PAPR is illustrated in Fig. 2. The simulation is run for 10^6 RCM-QAM blocks and each block includes 4096 RCM coded symbols. It is observed that, after passing through the shaping filter, the PAPR of RCM-QAM exceeds 10 dB at CCDF $= 10^{-4}$. In addition, the PAPR of RCM-QAM decreases with α increases, and RCM-QAM with $\alpha = 0.1$ offers about 0.7 dB higher PAPR than RCM-QAM with $\alpha = 0.4$ at CCDF $= 10^{-4}$.

2.2 The Proposed RCM-CPM

The proposed serially concatenated scheme of RCM-CPM represented by the diagram in Fig. 3. The high PAPR RCM coded symbols y_j are passed through a CPM modulator to obtain the 0 dB PAPR signal. The baseband signal output from CPM modulator is as follow

$$s(t) = \sqrt{\frac{E}{T}} e^{\sqrt{-1}\varphi(t)} \tag{4}$$

where E is the average symbol energy, T is the symbol duration and $\varphi(t)$ is the information-carrying phase given by

$$\varphi(t) = 2\pi h \sum_{j=0}^{\infty} y_j q(t - jT) \tag{5}$$

with

$$q(t) = \begin{cases} \int_0^t g(\tau)d\tau, t \leq LT \\ 1/2, t > LT \end{cases} \tag{6}$$

where y_j is the coded integer symbols of RCM and $g(t)$ is the frequency pulse, which can be chosen from rectangular, raised-cosine, Gaussian, *etc.* L is the frequency pulse length. Practically, $g(t)$ and the length L influence the smoothness of the signal. The modulation index h determines the magnitude of phase variation in a symbol interval. The smaller h, the higher bandwidth utilization, but at the same time, the higher BER. In the rest of this paper, we choose rectangular as $g(t)$ and set $L = 1$ to make the pulse shaping is just over a one-symbol interval, the so-called full response systems.

3 The Receiver

3.1 Overview

The RCM-CPM signal is assumed to be transmitted over an additive white Gaussian noise (AWGN) channel, the noisy signal to be processed at the receiver is

$$r(t) = s(t) + n(t) \tag{7}$$

where $n(t)$ denotes a complex white Gaussian random process with zero mean and variance $N_0/2$ per dimension.

As shown in Fig. 3, we employ two SISO modules at the receiver, BCJR algorithm [13] for CPM demodulating, and belief propagation (BP) algorithm [14] for RCM decoding. A serial iterative demodulating and decoding procedure is exploited to improve the performance of the system. The soft information is exchanged many times between BP decoder and BCJR demodulator. Since BP decoder of RCM is an iterative decoder, there are two iterative procedures to be performed in the soft information passing between BCJR demodulator and BP decoder: one iteration only performs BP iterative decoding, defined as the inner iteration, another one performs iteration between BCJR demodulating and BP decoding, defined as the outer iteration. We denote A as the maximum number of outer iterations, and B as the maximum number of inner iterations during each outer iteration.

The procedure of serial iterative demodulating and decoding procedure is described as follows. First, in a-th outer iteration, the BCJR demodulator computes the maximum a posterior (MAP) probability $P_{MAP}^{(a)}(y_j|r(t))$ based on the received symbol $r(t)$ and passes it to the BP decoder to update the probability of symbol nodes. Second, the BP decoder returns the updated probability of symbol nodes $P_j^{(a,B)}$ as the prior probability $P_{PP}^{(a)}(y_j)$ to the BCJR demodulator when inner iterations reach the maximum times B, then the $(a + 1)$-th outer iteration is started. Third, the hard decision is made by the BP decoder when the outer iterations reach A and inner iterations reach B.

3.2 BCJR Demodulating for CPM

The information-carrying phase during the j-th time interval, given in (5), can be rewritten when $L = 1$.

$$\varphi(t) = 2\pi h \sum_{j'=0}^{j} y_{j'} q(t - j'T)$$

$$= \pi h \sum_{j'=0}^{j-1} y_{j'} + 2\pi h y_j q(t - jT)$$

$$= \theta_j + \theta(t) \tag{8}$$

where θ_j is the accumulated phase and $\theta(t)$ is the incremental phase within one interval. From (8), we observe that the modulated signal over the j-th time interval depends both on the phase state denoted by accumulated phase θ_j and the most recent symbol y_j. For a rational modulation index $h = l/q$ (l and q are relatively prime integers), the phase evolution can be described by a finite-state machine, where each state is represented as $S_j = \{\theta_j\}$ and the number of such states is q or $2q$ respectively for l even or odd.

A SISO CPM demodulator is applied by using BCJR algorithm [13]. At every symbol interval $jT \le t \le (j + 1)T$, $(1 \le j \le M)$, BCJR achieves the MAP symbol decision that relies on the corresponding trellis state description of CPM. Denote the phase state set $\{\theta_j\}$, the MAP of the input y_j can be calculated by

$$P^{(a)}_{MAP}(y_j|r(t)) = \sum_{(\theta_j,\theta_{j+1})} P^{(a)}(\theta_j,\theta_{j+1}|r(t)) \tag{9}$$

where the superscript (a) represents the a-th iteration of BCJR, (θ_j,θ_{j+1}) represents a set of all possible state pair θ_j to θ_{j+1} at j-th symbol interval, and $\theta_{j+1} = \theta_j + \pi h y_j$. The probability of phase state varying from θ_j to θ_{j+1} can be factored as

$$P^{(a)}(\theta_j,\theta_{j+1}|r(t)) = \alpha^{(a)}_j(\theta_j)\gamma^{(a)}_j(\theta_j,\theta_{j+1})\beta^{(a)}_{j+1}(\theta_{j+1}) \tag{10}$$

where $\alpha^{(a)}_j(\theta_j)$ and $\beta^{(a)}_j(\theta_j)$ are computed respectively with forward and backward recursions as

$$\begin{cases} \alpha^{(a)}_j(\theta_j) = \sum\limits_{\theta_{j-1}} \alpha^{(a)}_{j-1}(\theta_{j-1})\gamma^{(a)}_{j-1}(\theta_{j-1},\theta_j) \\ \beta^{(a)}_j(\theta_j) = \sum\limits_{\theta_{j+1}} \beta^{(a)}_{j+1}(\theta_{j+1})\gamma^{(a)}_j(\theta_j,\theta_{j+1}) \end{cases} \tag{11}$$

The branch metric $\gamma^{(a)}_j(\theta_j,\theta_{j+1})$ under AWGN can be calculated by

$$\gamma^{(a)}_j(\theta_j,\theta_{j+1}) \propto \exp\left(\frac{\mathrm{Re}\left[\int_{jT}^{(j+1)T} r(t)s^*(t)dt\right]}{N_0}\right) P^{(a-1)}_{PP}(y_j) \tag{12}$$

where $s^*(t)$ represents the conjugation of $s(t)$, and the prior probability of phase state varying from θ_j to θ_{j+1} is represented by $P^{(a-1)}_{PP}(y_j)$. For the first outer iteration, $a = 1$, $P^{(0)}_{PP}(y_j)$ is set according to the prior PMF of RCM coded symbols, as given in Fig. 1. For the subsequent outer iteration, $a > 1$, $P^{(a-1)}_{PP}(y_j)$ is set to the PMF $P^{(a-1,B)}_j$ calculated by BP decoding.

3.3 BP Decoding for RCM

An RCM code can be represented by its corresponding bipartite graph, which consists in two sets of nodes: the variable nodes and the symbol nodes, the weighted connecting edge between one variable node and one symbol node corresponds to one entry of the mapping matrix. Hence, BP algorithm provides a powerful tool for RCM decoding which exchanges messages between these nodes.

Let us take a connecting pair of variable node x_i and symbol node y_j as an example to present the BP decoding algorithm for RCM, the weight on the connecting edge between them is w_{ij}. Denote N_j as the set of neighboring variable nodes of y_j and $N_j \backslash i$ as the set N_j excluding x_i. Denote M_i as the set of neighboring symbol nodes of x_i and $M_i \backslash j$ as the set M_i excluding y_j. Then, $y_j = \sum_{i' \in N_j} w_{i'j} x_{i'}$. Let $U^{(a,b)}_{ij}$ represent the message from x_i to y_j at a-th outer iteration and b-th inner iteration, and $V^{(a,b)}_{ji}$ represent the message from y_j to x_i. The BP decoding algorithm of RCM is described as follows.

(1) Initialization: For the first inner iteration, $a = 1$ and $b = 0$, the message sent from x_i to y_j is initialized with the a-priori probability.

$$\begin{cases} U_{ij}^{(1,0)}(0) = p(x_i = 0) = p_0 \\ U_{ij}^{(1,0)}(1) = p(x_i = 1) = 1 - p_0 \end{cases} \tag{13}$$

For the subsequent iteration, $a \geq 2$, the message sent from x_i to y_j is initialized by the outgoing messages from variable nodes of the last inner iteration.

$$\begin{cases} U_{ij}^{(a,0)}(0) = U_{ij}^{(a-1,B)}(0) \\ U_{ij}^{(a,0)}(1) = U_{ij}^{(a-1,B)}(1) \end{cases} \tag{14}$$

(2) Symbol Node Updating: Define $y_{j\backslash i} = \sum_{i' \in (\mathcal{N}_j \backslash i)} w_{i'j} x_{i'}$, its PMF at (a, b)-th iteration $P_{j\backslash i}^{(a,b)}$ is calculated by convolution

$$P_{j\backslash i}^{(a,b)} = (*)_{i' \in (\mathcal{N}_j \backslash i)} \left(P^{(a,b)}(w_{i'j} x_{i'}) \right) \tag{15}$$

where $(*)$ is the convolution of PMFs. The distribution of the weighted variables should be $P^{(a,b)}(w_{i'j} x_{i'} = 0) = U_{i'j}^{(a,b)}(0)$ and $P^{(a,b)}(w_{i'j} x_{i'} = w_{i'j}) = U_{i'j}^{(a,b)}(1)$. Then, $V_{ji}^{(a,b)}(0)$ and $V_{ji}^{(a,b)}(1)$ are computed based on $P_{j\backslash i}^{(a,b)}$ and $P_{MAP}^{(a,b)}(y_j | r(t))$ from BCJR as

$$\begin{cases} V_{ji}^{(a,b)}(0) = \sum_k P_{j\backslash i}(k) P_{MAP}^{(a,b)}(k | r(t)) \\ V_{ji}^{(a,b)}(1) = \sum_k P_{j|i}(k) P_{MAP}^{(a,b)}(k + w_{ij} | r(t)) \end{cases} \tag{16}$$

(3) Variable Node Updating: For x_i, compute $p_{i\backslash j}(0)$ and $p_{i\backslash j}(1)$ via multiplication

$$\begin{cases} p_{i\backslash j}(0) = p_0 \prod_{j' \in (\mathcal{M}_i \backslash j)} V_{j'i}^{(a,b)}(0) \\ p_{i\backslash j}(1) = (1 - p_0) \prod_{j' \in (\mathcal{M}_i \backslash j)} V_{j'i}^{(a,b)}(1) \end{cases} \tag{17}$$

Then, compute $U_{ij}^{(a,b)}$ via division and normalization

$$\begin{cases} U_{ij}^{(a,b)}(0) = \frac{p_{i\backslash j}(0)}{p_{i\backslash j}(0) + p_{i\backslash j}(1)} \\ U_{ij}^{(a,b)}(1) = 1 - U_{ij}^{(a,b)}(0) \end{cases} \tag{18}$$

(4) Output: After B inner iterations, BP stops. If the outer iteration does not reach A times, then calculate PMF $P_j^{(a,B)}$ by

$$P_j^{(a,B)} = (*)_{i' \in \mathcal{N}_j} \left(P^{(a,B)}(w_{i'j} x_{i'}) \right) \tag{19}$$

$P_j^{(a,B)}$ will be returned to BCJR as $P_{PP}^{(a)}(y_j)$ for the next outer iteration.

If the outer iteration reach the maximum times A, the iteration algorithm stops and makes a hard decision for x_i according to $p_i^{(A,B)}(0)$ and $p_i^{(A,B)}(1)$.

$$\hat{x}_i = \begin{cases} 0, & \text{if } p_i^{(A,B)}(0) > p_i^{(A,B)}(1) \\ 1, & \text{otherwise} \end{cases} \tag{20}$$

where $p_i^{(A,B)}(0) = p_0 \prod_{j' \in (\mathcal{M}_i)} V_{j'i}^{(A,B)}(0)$ and $p_i^{(A,B)}(1) = p_0 \cdot \prod_{j' \in (\mathcal{M}_i)} V_{j'i}^{(A,B)}(1)$.

3.4 The Performance of the Serial Iterative Demodulating and Decoding Procedure

In this sub-section, the simulation results are provided to evaluate the decoding performance of the proposed serial iterative demodulating and decoding procedure. We construct a 8192 × 4096 RCM mapping matrix with the weight set $\mathbf{W} = \{\pm 1, \pm 2, \pm 4, \pm 4\}$. Three combinations of outer iterations and inner iterations ($A = 1, B = 16$; $A = 2, B = 8$; $A = 4, B = 4$) are used for simulation. Thus, $C = A \times B = 16$ is a constant so as to ensure that these combinations have close computational complexity. In particular, the combination of $A = 1, B = 16$ means that the output messages of BP decoder does not return to BCJR again. Simulations are carried out under complex AWGN channel. Figure 4 shows BER performance of the proposed iterative demodulating and decoding procedure versus SNR at four different number of symbols. As can be observed from this figure, the different combination of A and B result in significantly different performance. The combination of $A = 2, B = 8$ achieves the best BER performance, which is at least 1 dB over that of $A = 1, B = 16$. This is because the inner BP iteration is approaching convergence at 8-th and that's the best time to return the message to outer iteration.

4 The Performance Comparison Between RCM-CPM and RCM-QAM

In this section, RCM-CPM is compared to the RCM-QAM in the presence of power amplified nonlinearities.

4.1 Power Amplifier Model

The input characteristic function $s_{in}(t)$ and output characteristic function $s_{out}(t)$ of nonlinear PA is written as below respectively:

$$s_{in}(t) = A_{in} \exp(j\phi_{in}(t))$$

$$s_{out}(t) = G(A_{in}(t)) \cdot \exp\left(j(\phi_{in}(t) + \Phi(A_{in}(t)))\right)$$

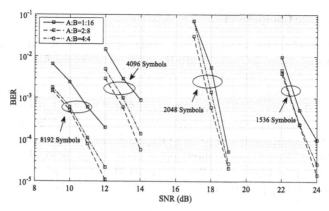

Fig. 4. The BER performance of RCM-CPM with different combinations of inner iterations and outer iterations

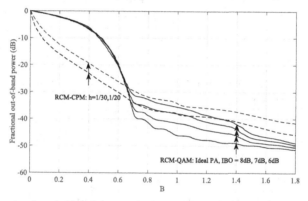

Fig. 5. The fractional out-of-band power of RCM-CPM versus RCM-QAM with nonlinear PA

In this paper, we use the solid-state power amplifier (SSPA) as the nonlinear PA [15]. The AM/AM conversion G for SSPA amplifier can be approximated by

$$G(A_{in}(t)) = \frac{A_{in}(t)}{\left(1 + \left(A_{in}(t)/A_{sat,in}\right)^{2p}\right)^{1/2p}}$$

and AM/PM conversions is

$$\Phi(A_{in}(t)) = 0$$

where $A_{sat,in}$ is the saturation level of the PA, and p controls the AM/AM sharpness of the saturation region. The PA nonlinearity is determined by the PAPR of the input signal and by the input back-off, defined as

$$IBO \equiv \frac{A_{sat,in}^2}{E\left(A_{in}^2(t)\right)} \geq 1 \tag{21}$$

In this paper, we set $p = 3$ for SSPA to simulate the performance of RCM-QAM.

4.2 Performance Comparison

We define the bandwidth that contains 99% of the signal power, which means that the fractional out-of-band power is -20 dB. Figure 5 compares the fraction out-of-band power of RCM-CPM and RCM-QAM with $\alpha = 0.4$. As can be seen from this figure, RCM-QAM is sensitive to the nonlinearity of PA, thus, sufficient power back-off is required to avoid spectral broadening. Constant RCM-CPM operates with 0 dB power back-off and the fractional out-of-band power depends on the modulation index h. Note that each constellation symbol in RCM-QAM carries two RCM coded symbols, but in RCM-CPM system one constellation symbol carries only one RCM coded symbols. Therefore, we set $h = 1/30$ so that RCMCPM and RCM-QAM have close bandwidth, as a result, the following performance comparison is performed under the same premise.

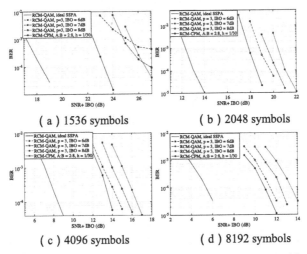

Fig. 6. The BER performance comparison of RCM-CPM and RCM-QAM with nonlinear power amplification

Figure 6 compares BER performance of RCM-CPM and RCM-QAM versus SNR at four different number of RCM coded symbols. The x-axis is adjusted to account for the negative impact of input power back-off. The BER of RCM-QAM is obtained in the presence of nonlinear PA with different IBO as 6 dB, 7 dB, 8 dB, respectively. Figure 6 demonstrates the advantage of RCM-CPM, which provides significant performance improvement over RCM-QAM due primarily to the 0 dB back-off. When the number of RCM coded symbols is 1536 at high SNRs, as shown in Fig. 6(a), RCM-QAMs have high error floors even IBO is set as high as 8 dB, RCM-CPM achieves the same BER with about 3 dB gain over RCM-QAM. For more coded symbols, as shown in Fig. 6(b)–(d), the IBO that results in the best RCM-QAM BER performance is 6 dB, which achieves closer performance to RCM-CPM as the number of symbols increases.

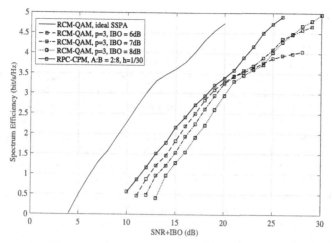

Fig. 7. The spectral efficiency comparison of RCM-CPM and RCM-QAM with nonlinear power amplification

Another primary evaluation metric of decoding performance for RCM is the spectral efficiency (in bits/s/Hz). For each block of source bits, the sender may keep generating and transmitting the constellation symbols with a given step size until an acknowledgment is received from receiver or the maximum number of transmission is reached. At the receiver, the RCM symbols are accumulated for decoding. If the decoded bits pass the cyclic redundancy check (CRC), an acknowledgment will be delivered to the sender, otherwise, increment symbols are needed to perform next round of decoding. Considering that RCM-CPM with $h = 1/30$ has almost the same bandwidth as RCM-QAM with $\alpha = 0.4$, therefore, the average the spectral efficiency of RCM is defined as the ratio of the number of correctly received source bits to the number of transmitted constellation symbols. The progressive transmission step is set to be $\Delta m = 64$ during simulation. The spectral efficiency of different RCM schemes are shown in Fig. 7. As can be observed from this figure, in the presence of nonlinear PA, the spectral efficiency of RCM-CPM is superior to RCM-QAMs with IBO. Especially when the spectral efficiency is higher than 4.5 at high SNRs, RCM-CPM outperforms RCM-QAMs with IBO by a factor of about 3 dB.

5 Conclusion

RCM reaches near-performance at the expense of high PAPR. This paper develops one feasible way to eliminate the PAPR problem associated RCM. In this method, the combination of RCM and CPM results in 0 dB PAPR constant envelope signals. This signal is ideally suited for nonlinear PA and could improve the efficiency of PA. At the receiver, the improved performance is achieved by employing the serial iterative demodulating and decoding procedure. It is demonstrated by simulation that RCM-CPM outperforms over RCM-QAM in terms of BER and spectral efficiency when the impact of nonlinear power amplification is taken into account, especially at high SNRs.

References

1. Cui, H., Luo, C., Wu, J., et al.: Compressive coded modulation for seamless rate adaptation. IEEE Trans. Wirel. Commun. **12**(10), 4892–4904 (2013)
2. Cui, H., Luo, C., Tan, K., et al.: Seamless rate adaptation for wireless networking. In: Proceedings of the 14th International Symposium on Modeling Analysis and Simulation of Wireless and Mobile Systems, MSWiM 2011, Miami, Florida, USA, 31 October–4 November 2011, DBLP (2011)
3. Duan, R., Liu, R., Shirvanimoghaddam, M., et al.: A low PAPR constellation mapping scheme for rate compatible modulation. IEEE Commun. Lett. **20**(2), 256–259 (2016)
4. Krongold, B.S., Jones, D.L.: An active-set approach for OFDM PAR reduction via tone reservation. IEEE Trans. Signal Process. **52**(2), 495–509 (2004)
5. Armstrong, J.: Peak-to-average power reduction for OFDM by repeated clipping and frequency domain filtering. Electron. Lett. **38**(5), 246 (2002)
6. Dandrea, A.N., Lottici, V., Reggiannini, R.: Nonlinear predistortion of OFDM signals over frequency-selective fading channels. IEEE Trans. Commun. **49**(5), 837–843 (2001)
7. Thompson, S.C., Ahmed, A.U., Proakis, J.G., et al.: Constant envelope OFDM. IEEE Trans. Commun. **56**(8), 1300–1312 (2008)
8. Wang, C., Cui, G., Wang, W.: Dual-stream transceiver structure with single antenna for phase-modulated OFDM. IEEE Commun. Lett. **20**(9), 1756–1759 (2016)
9. Cui, G., Wang, C., Wang, W.: Iterative detection with amplitude-phase demodulator for dual-stream CE-OFDM. IEEE Commun. Lett. **21**(9), 1 (2017)
10. Aulin, T., Sundberg, C.: Continuous phase modulation - part I: full response signaling. IEEE Trans. Commun. **29**(3), 196–209 (2003)
11. Narayanan, K.R., Altunbas, I., Narayanaswami, R.S.: Design of serial concatenated MSK schemes based on density evolution. IEEE Trans. Commun. **51**(8), 1283–1295 (2003)
12. Benaddi, T., Poulliat, C., Boucheret, M.L., et al.: Asymptotic analysis and design of LDPC codes for laurent-based optimal and suboptimal CPM receivers. In: IEEE International Conference on Acoustics. IEEE (2014)
13. Bahl, L.R., Cocke, J., Jelinek, F., et al.: Optimal decoding of linear codes for minimizing symbol error rate. IEEE Trans. Inf. Theory **20**(2), 284–287 (1974)
14. Lu, F., Dong, Y., Rao, W.: A parallel belief propagation decoding algorithm for rate compatible modulation. IEEE Commun. Lett. **21**(8), 1735–1738 (2017)
15. Rapp, C.: Effects of the HPA-nonlinearity on a 4-DPSK/OFDM signal for a digital sound broadcasting system. In: 2nd European Conference on Satellite Communications (ECSC) (1991)

Hybrid Precoding for HAP Massive MIMO Systems

Pingping Ji$^{(\boxtimes)}$, Lingge Jiang, Chen He, and Di He

Department of Electronic Engineering, Shanghai Jiao Tong University,
Shanghai 200240, China
{ji_pingping,lgjiang,chenhe,dihe}@sjtu.edu.cn

Abstract. A hybrid precoding scheme is proposed for high altitude platform (HAP) massive multiple-input multiple-output (MIMO) systems to obtain the radio frequency (RF) precoder and the baseband precoder with limited RF chains. We first exploit duality theory to derive the relation between the statistical channel state information (SCSI) and RF precoder, which is selected from a predefined codebook. Then, the baseband precoder is attained by zero forcing (ZF) based on the instantaneous effective channel matrix. A distinct performance gain is achieved by the proposed scheme according to the numerical results.

Keywords: High altitude platform · Massive MIMO · Rician fading channel · Hybrid precoding · Limited RF chains

1 Introduction

High altitude platforms (HAPs) with great advantages of low cost and flying on demand flexibly make an important contribution to the seamless integration of heterogeneous networks, such as terrestrial and satellite communication systems [1]. In next generation wireless communication system, higher data rate is demanded urgently and widely for people. The base station (BS) equipped with a large-scale antenna array, namely massive multiple-input multiple-output (MIMO), can achieve a significant increase in spectral efficiency [2]. Conventionally, digital precoding is applied to further make full use of multiple antennas, which means that each antenna requires a dedicated radio frequency (RF) chain. For massive MIMO, it is unrealistic due to its expensive cost in implementation. To handle this problem, hybrid precoding is proposed to obtain the RF precoder and the baseband precoder throughout this paper.

In time division duplexing (TDD) systems, channel station information at the transmitter (CSIT) can be directly acquired from uplink pilots due to the channel reciprocity, while in frequency division duplexing (FDD) systems, employed by most cellular systems today, the acquisition of accurate CSIT is challenging.

This work is supported by the National Natural Science Foundation of China under Grants 61771308, 91438113 and 61971278 and the Important National Science and Technology Specific Project of China under Grants 2018ZX03001020-005.

© Springer Nature Singapore Pte Ltd. 2020
Q. Yu (Ed.): SINC 2019, CCIS 1169, pp. 255–263, 2020.
https://doi.org/10.1007/978-981-15-3442-3_21

In [3,6], it was proved that the sum rate of massive MIMO can be improved greatly through hybrid precoding with accurate CSIT. In [3], a successive interference cancelation (SIC)-based hybrid precoding scheme is proposed by converting the non-convex constraints into a series of sub-problems, which reduces the complexity effectively. In [4], the RF precoder is obtained based on the practical phase constraints of RF chains and the zero-forcing (ZF) is applied to yield the baseband precoder. In [5], the RF precoder is selected from the codebook to minimize the distance between itself and the optimal precoder, i.e. the right singular vectors of the channel matrix, and the obtainment of the baseband precoder should satisfy the power constraint. In [6], a closed-form of mutual information is given and the lower bound of the sum rate is divided into two convex optimization problems to acquire the hybrid precoder.

In practical application of hybrid precoding, the acquisition of perfect channel state information (CSI) is properly unsuitable for HAP massive MIMO systems. The communication on HAP, which is at the altitude of 17–22 km, would have severe delay resulting from the long distance to its served users. In [7] and [8], the channel covariance is exploited and the signaling burden of CSI is reduced. In [7], the bi-convex problem is constructed by semidefinite relaxation to obtain the RF precoder based on the codebook. In [8], signal-to-leakage-plus-noise ratio (SLNR) is maximized to find the analog precoder. In [9] and [10], statistical CSI (SCSI) is utilized to lower the signaling cost. In [9], ergodic sum capacity is figured out by the duality theory and then the RF precoder is obtained based on the transformed convex problem. In [10], upper and lower bound of the average mutual information are derived and then the linear precoder is obtained by combining the left and right singular vectors.

In this paper, a hybrid precoding scheme is proposed to obtain the RF precoder and the baseband precoder with limited RF chains for HAP massive MIMO systems. First, the duality theory is applied to exploit the relation between SCSI and RF precoder. Then, the RF precoder, which is selected from a predefined codebook, is attained to maximize the capacity according to the obtained problem based on singular value decomposition (SVD). The baseband precoder is obtained by ZF based on the instantaneous effective channel matrix. Numerical results demonstrate that the proposed scheme achieves a better performance than the compared algorithms.

2   System Model

As illustrated in Fig. 1, we consider the downlink communication for HAP massive MIMO systems, where the HAP is equipped with an uniform planar antenna array (UPA) with N_T antennas, i.e. M antennas in each column and N antennas in each row ($N_T = M \times N$), and S RF chains to serve K single-antenna users such that $K \leq S < N_T$. The hybrid precoder is divided into a RF precoder and a baseband precoder as $\mathbf{FG}$, where $\mathbf{F} \in \mathbb{C}^{N_T \times S}$ is the RF precoder selected from a predefined codebook $\mathcal{F}$ and $\mathbf{G} \in \mathbb{C}^{S \times K}$ is the baseband precoder. Notably, the RF precoder $\mathbf{F}$ only changes phase, but the baseband precoder

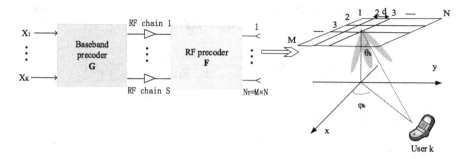

Fig. 1. System model with digital baseband precoding concatenated with analog RF precoding.

G modifies not only phase but also amplitude [4]. To lower the implementation cost, let the codebook $\mathcal{F}$ be discrete fourier transform (DFT) matrix as $\mathcal{F} \triangleq [\frac{1}{\sqrt{N_T}} e^{-\frac{j2\pi mn}{N_T}}]_{m,n=0,\cdots,N_T-1}$ [7].

The signal received at the users can be regarded as

$$\mathbf{y} = \mathbf{HFGx} + \mathbf{z} \tag{1}$$

where $\mathbf{x} \in \mathbb{C}^{K \times 1}$ is the transmitted data stream such that $\mathbb{E}[\mathbf{xx}^H] = \mathbf{I}_K$ where $\mathbb{E}[\cdot]$ is the expectation operator, $\mathbf{z} \sim \mathcal{CN}(0, \sigma^2 \mathbf{I}_K)$ is the additive white gaussian noise vector and the channel matrix is $\mathbf{H} \in \mathbb{C}^{K \times N_T}$ as

$$\mathbf{H} = \mathbf{\Phi}(\rho\bar{\mathbf{H}} + \sqrt{1 - \rho^2}\mathbf{H}_w) \tag{2}$$

where the diagonal element of $\mathbf{\Phi}$ represents the large-scale fading factor and $\mathbf{\Phi} = \text{diag}([\sqrt{(4\pi r_k/\lambda)^{-2}}]_{k=1,\cdots,K})$, where r_k is the distance between the HAP and the k-th user, λ is the wavelength and the operator $\text{diag}(\mathbf{a})$ is to make the elements of the vector a be the diagonal elements of a matrix, $\rho = \sqrt{K_f/(1 + K_f)}$ and K_f is the Rician factor, $\mathbf{H}_w$ is the Non-Line-of-Sight (NLOS) component and $(\mathbf{H}_w)_{ab} \sim \mathcal{CN}(0, 1)$ for $a = 1, \cdots, K$ and $b = 1, \cdots, N_T$, $\bar{\mathbf{H}} = [\bar{\mathbf{h}}_1, \bar{\mathbf{h}}_2, \cdots, \bar{\mathbf{h}}_K]^T$ is the Line-of-Sight (LOS) component and $\bar{\mathbf{h}}_k$ of the k-th user is given by [11]

$$\bar{\mathbf{h}}_k = \mathbf{a}(\theta_k, \varphi_k) \otimes \mathbf{b}(\theta_k, \varphi_k) \tag{3}$$

with

$$\mathbf{a}(\theta_k, \varphi_k) = [1, e^{-j2\pi d \sin\theta_k \cos\varphi_k/\lambda}, \cdots, \\ e^{-j2\pi(M-1)d \sin\theta_k \cos\varphi_k/\lambda}] \tag{4}$$

and

$$\mathbf{b}(\theta_k, \varphi_k) = [1, e^{-j2\pi d \sin\theta_k \sin\varphi_k/\lambda}, \cdots, \\ e^{-j2\pi(N-1)d \sin\theta_k \sin\varphi_k/\lambda}] \tag{5}$$

where $\theta_k \in [0, \pi/2]$ and $\varphi \in [-\pi, \pi]$ are the vertical and the horizontal angle of departure (AoD) of the k-th user respectively, $d = 2\lambda$ is the antenna spacing.

For given channel matrix $\mathbf{H}$ and given hybrid precoder $(\mathbf{F}, \mathbf{G})$, we can achieve the maximum sum rate with effective channel $\mathbf{HF}$ as

$$R(\mathbf{F}, \mathbf{G}, \mathbf{H}) = \sum_{k=1}^{K} \log_2(1 + \frac{|\mathbf{h}_k^H \mathbf{Fg}_k|^2}{\sum_{l>k} \mathbf{h}_k^H \mathbf{Fg}_l \mathbf{g}_l^H \mathbf{F}^H \mathbf{h}_k + \sigma^2}) \tag{6}$$

3   Hybrid Precoding Design

The ergodic sum capacity is

$$C(\mathbf{H}) = \max_{\mathbf{F} \in \mathcal{F}} \mathbb{E}[\max_{\mathbf{G}: \mathrm{Tr}(\mathbf{FGG}^H \mathbf{F}^H) \leq P} R(\mathbf{F}, \mathbf{G}, \mathbf{H})] \tag{7}$$

where $\mathrm{Tr}(\mathbf{FGG}^H \mathbf{F}^H) \leq P$ is set to satisfy the power constraint and P is the total power.

We let the power allocated to k-th user as $p_k = P/K$ and power allocation matrix $\mathbf{P} = \mathrm{diag}(p_1, \cdots, p_K)$. According to the duality theory [9], the ergodic sum capacity can be stated as

$$C(\mathbf{H}) = \max_{\mathbf{F} \in \mathcal{F}} \mathbb{E}[\log_2 |\mathbf{I}_S + \frac{1}{\sigma^2} \mathbf{F}^H \mathbf{H}^H \mathbf{PHF}|] \tag{8}$$

Theorem 1 ([13], *Theorem 7.7.1*). *Let* $\mathbf{A} \in \mathbb{C}^{K \times K} > 0$. *Then*

$$\mathbb{E}[\mathbf{H}_w^H \mathbf{A} \mathbf{H}_w] = \mathrm{Tr}(\mathbf{A}) \mathbf{I}_{N_T} \tag{9}$$

Combing with (2), the optimization function in (8) can be derived as

$$
\begin{aligned}
&\mathbb{E}[\log_2 |\mathbf{I}_S + \frac{1}{\sigma^2} \mathbf{F}^H \mathbf{H}^H \mathbf{PHF}|] \\
&= \mathbb{E}[\log_2 |\mathbf{I}_S + \frac{1}{\sigma^2} \mathbf{F}^H (\rho \bar{\mathbf{H}}^H + \sqrt{1-\rho^2} \mathbf{H}_w^H) \mathbf{\Phi}^H \mathbf{P \Phi} \\
&\qquad (\rho \bar{\mathbf{H}} + \sqrt{1-\rho^2} \mathbf{H}_w) \mathbf{F}|] \\
&\overset{(a)}{\leq} \log_2 |\mathbf{I}_S + \frac{\rho^2}{\sigma^2} \mathbf{F}^H \bar{\mathbf{H}}^H \mathbf{\Phi}^H \mathbf{P \Phi} \bar{\mathbf{H}} \mathbf{F} \\
&\qquad + \frac{1-\rho^2}{\sigma^2} \mathbf{F}^H \mathbb{E}[\mathbf{H}_w^H \mathbf{\Phi}^H \mathbf{P \Phi} \mathbf{H}_w] \mathbf{F}| \\
&\overset{(b)}{=} \log_2 |\mathbf{I}_S + \frac{\rho^2}{\sigma^2} \mathbf{F}^H \bar{\mathbf{H}}^H \mathbf{\Phi}^H \mathbf{P \Phi} \bar{\mathbf{H}} \mathbf{F} \\
&\qquad + \frac{1-\rho^2}{\sigma^2} \mathrm{Tr}(\mathbf{\Phi}^H \mathbf{P \Phi}) \mathbf{F}^H \mathbf{F}| \\
&\overset{a.s.}{\longrightarrow} \log_2 |w \mathbf{I}_S + \frac{\rho^2}{\sigma^2} \mathbf{F}^H \bar{\mathbf{H}}^H \mathbf{\Phi}^H \mathbf{P \Phi} \bar{\mathbf{H}} \mathbf{F}|
\end{aligned} \tag{10}
$$

where (a) follows Jensen's inequality, (b) follows (9) in Theorem 1, $\overset{a.s.}{\longrightarrow}$ denotes almost sure convergence, such that $\mathbf{F}^H \mathbf{F} \overset{a.s.}{\longrightarrow} \mathbf{I}_S$ under asymptotic setting $N_T \rightarrow \infty$, and $w = 1 + \frac{1-\rho^2}{\sigma^2} \mathrm{Tr}(\mathbf{\Phi}^H \mathbf{P \Phi})$.

Algorithm 1. RF Precoding Algorithm for Solving Problem (12)

Require: $\mathbf{V}_1$

1: **Step 1:** $\boldsymbol{\Psi} = \mathbf{C}^H \mathbf{V}_1$
2: **Step 2:** Let $\mathbf{d} \in \mathbb{C}^{N_T \times 1}$ be a binary vector as 0 or 1 of which entry one elements are all at the location of the S largest elements from the diagonal elements ($\boldsymbol{\Psi}\boldsymbol{\Psi}^H$).
3: **Step 3:** The RF precoder $\mathbf{F} \in \mathbb{C}^{N_T \times S}$ consists of the nonzero columns of the matrix $\mathbf{C} \times \mathrm{diag}(\mathbf{d})$.

Ensure: RF precoder $\mathbf{F}$

According to (10), the ergodic sum capacity optimization problem by designing the RF precoder $\mathbf{F}$ can be stated as

$$\mathbf{F}^{\mathrm{opt}} = \arg\max_{\mathbf{F} \in \mathcal{F}} \; \log_2 |w\mathbf{I}_S + \frac{\rho^2}{\sigma^2}\mathbf{F}^H\bar{\mathbf{H}}^H\boldsymbol{\Phi}^H\mathbf{P}\boldsymbol{\Phi}\bar{\mathbf{H}}\mathbf{F}| \tag{11}$$

Let the SVD of $\mathbf{P}^{1/2}\boldsymbol{\Phi}\bar{\mathbf{H}}$ as $\mathbf{P}^{1/2}\boldsymbol{\Phi}\bar{\mathbf{H}} = \mathbf{U}\boldsymbol{\Sigma}\mathbf{V}^H$, where $\mathbf{U}$ is a $K \times K$ unitary matrix, $\boldsymbol{\Sigma} = [\boldsymbol{\Sigma}_K \; \mathbf{0}_{K \times (N_T-K)}]$, $\boldsymbol{\Sigma}_K$ is a $K \times K$ diagonal matrix of singular values in a decreasing order, and $\mathbf{V}$ is an $N_T \times N_T$ unitary matrix. From the definition of $\boldsymbol{\Phi}$ and $\mathbf{P}$, $\mathbf{P}^{1/2}\boldsymbol{\Phi}$ is a diagonal matrix. In particular, we define $\boldsymbol{\Sigma} \triangleq [\boldsymbol{\Sigma}_1 \; \mathbf{0}_{K \times (N_T-S)}]$, where $\boldsymbol{\Sigma}_1 = [\boldsymbol{\Sigma}_K \; \mathbf{0}_{K \times (S-K)}]$ is of dimension $K \times S$, and $\mathbf{V} \triangleq [\mathbf{V}_1 \; \mathbf{V}_2]$, where $\mathbf{V}_1$ is of dimension $N_T \times S$.

Proposition 1. *The optimization problem (11)*

$$\mathbf{F}^{\mathrm{opt}} = \arg\max_{\mathbf{F} \in \mathcal{F}} \; \log_2 |w\mathbf{I}_S + \frac{\rho^2}{\sigma^2}\mathbf{F}^H\bar{\mathbf{H}}^H\boldsymbol{\Phi}^H\mathbf{P}\boldsymbol{\Phi}\bar{\mathbf{H}}\mathbf{F}|$$

is equivalent to the following problem

$$\mathbf{F}^{\mathrm{opt}} = \arg\min_{\mathbf{F} \in \mathcal{F}} \; \|\mathbf{V}_1 - \mathbf{F}\|_F \tag{12}$$

Proof. The optimization function in (11) can be further simplified as

$$\log_2 |w\mathbf{I}_S + \frac{\rho^2}{\sigma^2}\mathbf{F}^H\bar{\mathbf{H}}^H\boldsymbol{\Phi}^H\mathbf{P}\boldsymbol{\Phi}\bar{\mathbf{H}}\mathbf{F}|$$

$$= \log_2(w^S) + \log_2 |\mathbf{I}_S + \frac{\rho^2}{w\sigma^2}\mathbf{F}^H\bar{\mathbf{H}}^H\boldsymbol{\Phi}^H\mathbf{P}\boldsymbol{\Phi}\bar{\mathbf{H}}\mathbf{F}|$$

$$= \log_2(w^S) + \log_2 |\mathbf{I}_S + \frac{\rho^2}{w\sigma^2}\mathbf{F}^H\mathbf{V}\boldsymbol{\Sigma}^H\boldsymbol{\Sigma}\mathbf{V}^H\mathbf{F}|$$

$$= \log_2(w^S) + \log_2 |\mathbf{I}_S + \frac{\rho^2}{w\sigma^2}\mathbf{F}^H[\mathbf{V}_1 \; \mathbf{V}_2]\begin{bmatrix} \boldsymbol{\Sigma}_1^H \\ \mathbf{0}_{(N_T-S) \times K} \end{bmatrix}$$

$$[\boldsymbol{\Sigma}_1 \; \mathbf{0}_{K \times (N_T-S)}]\begin{bmatrix} \mathbf{V}_1^H \\ \mathbf{V}_2^H \end{bmatrix}\mathbf{F}| \tag{13}$$

$$= \log_2(w^S) + \log_2 \left| \mathbf{I}_S + \frac{\rho^2}{w\sigma^2} \mathbf{F}^H \mathbf{V}_1 \mathbf{\Sigma}_1^H \mathbf{\Sigma}_1 \mathbf{V}_1^H \mathbf{F} \right|$$

$$\overset{(a)}{=} \log_2(w^S) + \log_2 \left| \mathbf{I}_S + \frac{\rho^2}{w\sigma^2} \mathbf{\Sigma}_1^2 \mathbf{V}_1^H \mathbf{F} \mathbf{F}^H \mathbf{V}_1 \right|$$

$$\overset{(b)}{=} \log_2(w^S) + \log_2 \left| \mathbf{I}_S + \frac{\rho^2}{w\sigma^2} \mathbf{\Sigma}_1^2 \right| + \log_2 \left| \mathbf{I}_S \right.$$

$$- (\mathbf{I}_S + \frac{\rho^2}{w\sigma^2} \mathbf{\Sigma}_1^2)^{-1} \times \frac{\rho^2}{w\sigma^2} \mathbf{\Sigma}_1^2 (\mathbf{I}_S - \mathbf{V}_1^H \mathbf{F} \mathbf{F}^H \mathbf{V}_1) \left. \right|$$

$$\overset{(c)}{\approx} \log_2(w^S) + \log_2 \left| \mathbf{I}_S + \frac{\rho^2}{w\sigma^2} \mathbf{\Sigma}_1^2 \right| + \log_2 \left| \mathbf{V}_1^H \mathbf{F} \mathbf{F}^H \mathbf{V}_1 \right|$$

where (a) follows the determinant theorem of Sylvester, such as $|\mathbf{I} + \mathbf{AB}| = |\mathbf{I} + \mathbf{BA}|$, (b) follows the fact that $\mathbf{I} + \mathbf{AB} = (\mathbf{I} + \mathbf{A})(\mathbf{I} - (\mathbf{I} + \mathbf{A})^{-1}\mathbf{A}(\mathbf{I} - \mathbf{B}))$, and (c) follows from adopting a high Rician factor K_f, i.e. $\rho = \frac{K_f}{1+K_f}$, which implies that $(\mathbf{I}_S + \frac{\rho^2}{w\sigma^2}\mathbf{\Sigma}_1^2)^{-1}\frac{\rho^2}{w\sigma^2}\mathbf{\Sigma}_1^2 \approx \mathbf{I}_S$, because $w = 1 + \frac{1-\rho^2}{\sigma^2}\text{Tr}(\mathbf{\Phi}^H \mathbf{P} \mathbf{\Phi})$ and $\frac{\rho^2}{w\sigma^2} = \frac{\rho^2}{\sigma^2 + (1-\rho^2)\text{Tr}(\mathbf{\Phi}^H \mathbf{P} \mathbf{\Phi})}$.

From (13), we observe that designing $\mathbf{F}$ is to find the largest projection on $\mathbf{V}_1$, which is equal to find the smallest Euclidean distance between $\mathbf{V}_1$ and $\mathbf{F}$, i.e.

$$\mathbf{F}^{\text{opt}} = \underset{\mathbf{F} \in \mathcal{F}}{\arg \min} \| \mathbf{V}_1 - \mathbf{F} \|_F.$$

∎

Since $\mathbf{F} \in \mathcal{F}$ and $\mathcal{F}$ is DFT matrix as $\mathcal{F} \triangleq [\frac{1}{\sqrt{N_T}} e^{-\frac{j2\pi mn}{N_T}}]_{m,n=0,\cdots,N_T-1}$, the obtainment of RF precoder $\mathbf{F}$ is to find the matrix belonging to $\mathcal{F}$ where $\mathbf{F}$ has the maximum projection on $\mathbf{V}_1$.

To maximize the ergodic sum capacity in (8), we propose an RF precoding algorithm in Algorithm 1. Let the matrix consisting of the vectors in $\mathcal{F}$ as $\mathbf{C} \in \mathbb{C}^{N_T \times N_T}$, where the amplitude of each column of $\mathbf{C}(j)$ is 1. Notice that for any feasible solution $\mathbf{F} \in \mathcal{F}$, $\mathbf{F}$ is composed of the nonzero columns of $\mathbf{C} \times \text{diag}(\mathbf{d})$, where the temporary vector $\mathbf{d} = [d_1, \cdots, d_{N_T}] \in \{0,1\}^{N_T}$ and $\|\mathbf{d}_0\| = S$. $\mathbf{d}$ is a selection matrix form the matrix $\mathbf{C}$ by selecting the largest projection on $\mathbf{V}_1$.

Remark 1. We observe that the RF precoder $\mathbf{F}$ only requires parameters $\mathbf{V}_1$ and codebook $\mathbf{C}$, which can be obtained at the HAP. The HAP has knowledge of the users distribution and slowly varying LOS component, instead of instantaneous CSI. Therefore, the proposed RF precoding algorithm is suitable for HAP massive MIMO systems.

4   Numerical Results and Analysis

In this section, the performance of the proposed hybrid precoding algorithm is evaluated. We assume that the height of HAP is 20 km, the HAP is equipped with UPA $N_T = \sqrt{M} = \sqrt{N} = 100$, the users are randomly distributed below the HAP in a circle whose radius is $20\,\mathrm{km}$. We set the frequency and bandwidth as $2.4\,\mathrm{GHz}$ and $10\,\mathrm{MHz}$ respectively. The noise variance is -169 dBm/Hz.

We combine the proposed RF precoding algorithm with simple ZF baseband precoding. We compare the proposed RF algorithm with two algorithms: one is low complexity (LC) + ZF precoding in [4] and the other is semidefinite relaxation (SDR) + ZF precoding in [9]. In [4], LC algorithm obtains RF precoder as $\mathbf{F}_{i,j} = \frac{1}{\sqrt{N_T}} e^{j\eta_{i,j}}$, where $\eta_{i,j}$ is the phase of the $(i.j)$th element of the matrix $(\sqrt{P}\boldsymbol{\Phi}\bar{\mathbf{H}})^H$ as shown in (11). In [9], SDR algorithm derives RF precoder $\mathbf{F}$ consisting of the nonzero columns of $\mathbf{C} \times \mathrm{diag}(\mathbf{d})$, where $\mathbf{D} = \mathrm{diag}(\mathbf{d})$ can be obtained from the convex problem as

$$\mathbf{D} = \underset{\mathbf{d}\in\mathbb{R}^{N_T},\mathbf{D}-\mathbf{D}^2\succeq 0,\mathrm{Tr}(\mathbf{D})=S}{\arg\max} \log_2 |\mathbf{I}_{N_T} + \mathbf{ED}| \tag{14}$$

where $\mathbf{E} = \frac{\rho^2}{w\sigma^2}\mathbf{C}^H\bar{\mathbf{H}}^H\boldsymbol{\Phi}^H\mathbf{P}\boldsymbol{\Phi}\bar{\mathbf{H}}\mathbf{C}$, and (14) is derived from (11) based on semidefinite relaxation as

$$\log_2 |\mathbf{I}_S + \tfrac{\rho^2}{w\sigma^2}\mathbf{F}^H\bar{\mathbf{H}}^H\boldsymbol{\Phi}^H\mathbf{P}\boldsymbol{\Phi}\bar{\mathbf{H}}\mathbf{F}|$$
$$= \log_2 |\mathbf{I}_{N_T} + \tfrac{\rho^2}{w\sigma^2}\bar{\mathbf{H}}^H\boldsymbol{\Phi}^H\mathbf{P}\boldsymbol{\Phi}\bar{\mathbf{H}}\mathbf{F}\mathbf{F}^H|$$
$$= \log_2 |\mathbf{I}_{N_T} + \tfrac{\rho^2}{w\sigma^2}\bar{\mathbf{H}}^H\boldsymbol{\Phi}^H\mathbf{P}\boldsymbol{\Phi}\bar{\mathbf{H}}\mathbf{C}\mathbf{D}\mathbf{C}^H| \tag{15}$$
$$= \log_2 |\mathbf{I}_{N_T} + \tfrac{\rho^2}{w\sigma^2}\mathbf{C}^H\bar{\mathbf{H}}^H\boldsymbol{\Phi}^H\mathbf{P}\boldsymbol{\Phi}\bar{\mathbf{H}}\mathbf{C}\mathbf{D}|$$
$$= \log_2 |\mathbf{I}_{N_T} + \mathbf{ED}|$$

where the elements of vector $\mathbf{d}$ is binary, such as $\mathbf{D}-\mathbf{D}^2 = 0$, where the constraint is non-convex, and then we apply the semidefinite relaxation $\mathbf{D} - \mathbf{D}^2 \succeq 0$. SDR algorithm obtains RF precoder by solving problem (14).

Remark 2. We observe that LC algorithm requires the same number of RF chains S as the number of users K, while the proposed algorithm and SDR algorithm do not have this limitation, just with the limitation of $K \leq S$.

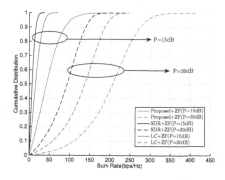

Fig. 2. Cumulative distribution of the sum rate by different transmit power P with $K = 40$ users, $S = 40$ RF chains, and Rician factor $K_f = 10$ dB.

As shown in Fig. 2, the 10000 randomly generated samples is taken to yield the cumulative distribution of the sum rate and it can be concluded that the convergence of Algorithm 1 is preserved by the ascending shape of the curves.

Figure 3 depicts the performance of the proposed hybrid precoding algorithm compared with LC + ZF scheme in [4] and SDR + ZF scheme in [9] in terms of transmit power. We can see that the sum rate of the proposed algorithm is higher than that of the other two comparisons as shown in Figs. 2 and 3. The sum rate grows stably with the increase of the transmit power. As the proposed RF precoding algorithm applies the SVD, which is better than LC and SDR algorithm explicitly. The SDR algorithm performs the worst due to the relaxation of $\mathbf{D} - \mathbf{D}^2 \succeq 0$.

Figure 4 depicts the performance of the proposed hybrid precoding algorithm compared with LC + ZF scheme in [4] as a function of the number of RF chains. The proposed outperforms the LC + ZF with the same reasons as illustrated above. The sum rate grows stably with the increase of the number of RF chains. It can be concluded that the sum rate will achieve an upper limit when S grows approaching N_T, which is consistent with the result in [9].

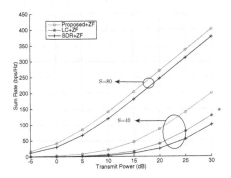

Fig. 3. The sum rate versus transmit power P by different RF chains S with $K = 40$ users and Rician factor $K_f = 10\,dB$.

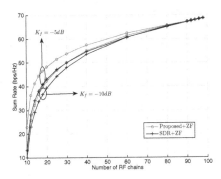

Fig. 4. The sum rate versus the number of RF chains S by different Rician factor K_f with $K = 10$ users and transmit power $P = 10\,dB$.

5 Conclusion

In this paper, we have proposed a hybrid precoding scheme, which consists of a RF precoder and a baseband precoder, to limit the RF chains for HAP massive MIMO systems. The duality theory has been applied to exploit the ergodic sum capacity in Rician fading channel, which only depends on the LOS component, i.e. SCSI. Then, the RF precoder has been obtained based on SVD. The baseband precoder has been attained through the instantaneous effective channel matrix by ZF. Numerical results have demonstrated that the hybrid precoding algorithm outperforms the compared algorithms.

References

1. Mohammed, A., Mehmood, A., Pavlidou, F., Mohorcic, M.: The role of High-Altitude Platforms (HAPs) in the global wireless connectivity. Proc. IEEE **99**(11), 1939–1953 (2011)
2. Marzetta, T.L.: Noncooperative cellular wireless with unlimited numbers of base station antennas. IEEE Trans. Wireless Commun. **9**(11), 3590–3600 (2010)
3. Gao, X., Dai, L., Han, S., Chih-Lin, I., Heath, R.W.: Energy-efficient hybrid analog and digital precoding for MmWave MIMO systems with large antenna arrays. IEEE J. Sel. Areas Commun. **34**(4), 998–1009 (2016)
4. Liang, L., Xu, W., Dong, X.: Low-complexity hybrid precoding in massive multiuser MIMO systems. IEEE Wirel. Commun. Lett. **3**(6), 653–656 (2014)
5. Ayach, O.E., Rajagopal, S., Abu-Surra, S., Pi, Z., Heath, R.W.: Spatially sparse precoding in millimeter wave MIMO systems. IEEE Trans. Wirel. Commun. **13**(3), 1499–1513 (2014)
6. He, L., Wang, J., Song, J.: Spatial modulation for more spatial multiplexing: RF-chain-limited generalized spatial modulation aided MM-Wave MIMO with hybrid precoding. IEEE Trans. Commun. **66**(3), 986–998 (2018)
7. Liu, A., Lau, V.: Phase only RF precoding for massive MIMO systems with limited RF chains. IEEE Trans. Sign. Proces. **62**(17), 4505–4515 (2014)
8. Park, S., Park, J., Yazdan, A., Heath, R.W.: Exploiting spatial channel covariance for hybrid precoding in massive MIMO systems. IEEE Trans. Sign. Process. **65**(14), 3818–3832 (2017)
9. Liu, A., Lau, V.K.N.: Impact of CSI knowledge on the codebook-based hybrid beamforming in massive MIMO. IEEE Trans. Sign. Process. **64**(24), 6545–6556 (2016)
10. Zeng, W., Xiao, C., Wang, M., Lu, J.: Linear precoding for finite-alphabet inputs over MIMO fading channels with statistical CSI. IEEE Trans. Sign. Process. **60**(6), 3134–3148 (2012)
11. Xu, Y., Xia, X., Xu, K., Wang, Y.: Three-dimension massive MIMO for air-to-ground transmission: location-assisted precoding and impact of AoD uncertainty. IEEE Access **5**, 15582–15596 (2017)
12. Zhang, Q., Xi, Q., He, C., Jiang, L.: User clustered opportunistic beamforming for stratospheric communications. IEEE Commun. Lett. **20**(9), 1832–1835 (2016)
13. Gupta, A.K., Nagar, D.K.: Matrix Variate Distributions. Chapman, Hall/CRC, Boca Raton (2000)

The Approach to Satellite Anti-interception Communication Based on WFRFT-TDCS

Yuan Qiu[✉], Haiyu Ren, Longfei Gao, and Yichen Xiao

Institute of Software, Chinese Academy of Sciences, Beijing 100190, China
qiuyuan@iscas.ac.cn

Abstract. In order to improve the security and concealment of satellite communication, an anti-interception technology based on weighted fractional Fourier transform (WFRFT) and transform domain communication system (TDCS) is proposed. WFRFT replaces FFT/IFFT and CCSK frequency-domain mapping replaces time-domain modulation in TDCS. Moreover, the receiving basis function is designed according to the principle of equal gain combining. The simulation results show that, WFRFT-TDCS anti-eavesdropping technology can be applied to satellite transceiver of spectrum mismatch, as well as maintain system complexity and enhance anti-scanning performance. Even if the SNR is greatly increased, unauthorized receiver cannot demodulate the signal correctly. For authorized receiver, SNR loss is less than 1 dB when the transmission spectrum availability is more than 80%.

Keywords: Transform domain communication system · Weighted fractional Fourier transform · Anti-interception communication · Spectrum mismatch · Satellite communication

1 Introduction

Compared with mobile communication networks, satellite networks are not constrained by geographical conditions and have a wide range of signal coverage.

The above advantages are based on the premise of ensuring satellite networks' security. Because satellite networks are often deployed in the whole space area, satellite communication (SATCOM) is insecure due to the broadcast nature of radio propagation. This leads to poor signal concealment and confidential information leakage. Furthermore, there is eavesdroppers' threat in both satellite-ground channel and ground-ground channel, which increases the risk of eavesdropping. Therefore, anti-interception capability of SATCOM is attracting considerable attentions recently.

In order to cope with the rapid development of signal identification and parametric estimation technology, new anti-eavesdropping methods have emerged, such as chaotic technology, cognitive radio (CR), transform domain communication system (TDCS), weighted fractional Fourier transform (WFRFT), etc.

This work is supported by the National Key Research and Development Program of China (No. 2016YFB0501104).

© Springer Nature Singapore Pte Ltd. 2020
Q. Yu (Ed.): SINC 2019, CCIS 1169, pp. 264–274, 2020.
https://doi.org/10.1007/978-981-15-3442-3_22

TDCS [1, 2] is a CR-based intelligent radio technology, which dynamically generates the basis function (BF) in the transform domain for avoiding interference spectrum. Note that anti-interception characteristic of traditional TDCS depends on the pseudo-random phase sequence design in the BF [3, 4]. In recent years, the researches on TDCS at home and abroad mostly take the consistency of transceiver spectrum as the precondition [5, 6]. In SATCOM system, the differences of geographic location and electromagnetic environment between transceivers do not satisfy the above assumption, thus deteriorating system performance. Hence the application of TDCS in SATCOM needs to overcome the problem of spectrum mismatch at first.

WFRFT [7–10] is a new signal processing method, whose baseband constellation has the characteristics of fuzziness and fission by the rational design of weighted parameters. Its Gauss-like statistical properties has unique advantages in anti-scanning communication. However, the classical 4-WFRFT has only one weighted parameter, so eavesdroppers can intercept original signal by scanning modulation order at the cost of time.

In secure communication, the combinatory strategy to embed WFRFT signals into TDCS is introduced in [11–14], which is tempting to make full use of the properties of TDCS and WFRFT.

To improve safety and reliability of SATCOM from various angles, an anti-interception communication approach based on WFRFT-TDCS is proposed. TDCS achieves anti-eavesdropping capacity by tamed spread spectrum over non-interference frequency band. Additionally, aiming at solving the problem of spectrum inconsistency, the receiving BF design is introduced, based upon equal gain combining (EGC) principle. Furthermore, WFRFT is used to replace FFT/IFFT in TDCS. On the premise of maintaining system complexity, it changes the spatial distribution of TDCS signals and improves the anti-scanning ability.

The remainder of this paper is organized as follows: Section 2 presents the definitions and properties of WFRFT. In Section 3, the system model of WFRFT-TDCS is described and anti-interception communication mechanism is given. The BER curves are simulated and analyzed in Section 4. Finally, Section 5 concludes this paper.

2 Weighted Fractional Fourier Transform

WFRFT is a promising technique for signal processing. In this section, we introduce its basic definitions and investigate its anti-parameter scanning properties.

2.1 Definition

Given an arbitrary complex vector $X_0(n)$, the α-order WFRFT is defined by

$$
\begin{aligned}
S_0(n) &= F^\alpha[X_0(n)] \\
&= \omega_0(\alpha)X_0(n) + \omega_1(\alpha)X_1(n) \\
&\quad + \omega_2(\alpha)X_2(n) + \omega_3(\alpha)X_3(n)
\end{aligned}
\tag{1}
$$

where F^α denotes the WFRFT kernel (or basic operator). $X_l(n)(l = 0, 1, 2, 3)$ represents 0– 3 times of FFT operation results respectively. The weighted coefficients $\omega_l(\alpha)(l = 0, 1, 2, 3)$ is given by

$$\omega_l(\alpha) = \frac{1}{4} \sum_{k=0}^{3} \exp\left[\pm\frac{2\pi i}{4}(\alpha - l)k\right], \ (l = 0, 1, 2, 3) \tag{2}$$

where α is chosen in the real interval [0,4]. According to the symmetry of FFT, WFRFT process is implemented through one N-point FFT module and two reverse modules in Fig. 1. The physical expression is generalized as

$$S_0(n) = \omega_0(\alpha)X_0(n) + \omega_2(\alpha)X_0(N - n)$$
$$+ \omega_1(\alpha)\frac{1}{\sqrt{N}} \sum_{k=0}^{N-1} X_0(k) \exp\left(j\frac{2\pi kn}{N}\right)$$
$$+ \omega_3(\alpha)\frac{1}{\sqrt{N}} \sum_{k=0}^{N-1} X_0(N - k) \exp\left(-j\frac{2\pi kn}{N}\right) \tag{3}$$

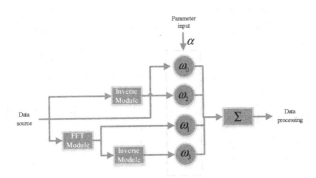

Fig. 1. WFRFT physical implementation

2.2 Anti-interception Analysis

By varying the weighted coefficients, WFRFT make transformed signals present Gauss-like distribution on the complex plane. Only by grasping weighted coefficients and implementing inverse transform can the constellation be restored. The analysis of Gauss-like feature and constellation splitting is discussed in this section. Specifically, anti-eavesdropping abilities can be studied by the qualitative analysis of the detection and modulation recognition probability.

Because both FFT and invert FFT are linear transforms, it is known from the central limit theorem that when multiple random variables with the same distribution are linearly combined, their synthetic variables have Gauss-like characteristic. The linear combination of frequency-domain term $X_1(n)$ and $X_3(n)$ constitutes the Gauss-like function $F^\alpha(\cdot)$

$$F^\alpha(X_0(n)) = \omega_1(\alpha)X_1(n) + \omega_3(\alpha)X_3(n) \tag{4}$$

whose amplitude directly reflects the power of Gauss-like noise.

The Gauss-like characteristic of WFRFT in Fig. 2 ensures the invalidity of conventional energy-based and periodic feature-based detection, thus ensuring the low detection probability.

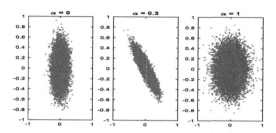

Fig. 2. The Gauss-like function of BPSK

In contrast to the above case, when the time-domain terms $X_0(n)$ and $X_2(n)$ are synthesized, the uncertainty of original data and the variability of the weighting coefficients show rich combination patterns. This time-domain term synthesis is attributed to the splitting of signal constellation. The constellation fission function $T^\alpha(\cdot)$ is defined by

$$T^\alpha(X_0(n)) = \omega_0(\alpha)X_0(n) + \omega_2(\alpha)X_2(n) \tag{5}$$

Generally speaking, on the basis of specific modulation, constellation points are relatively fixed on the constellation diagram. However, constellation points change with the change of WFRFT weighted coefficients, as shown in Fig. 3. Thus, effective constellation camouflage can be realized to guarantee low modulation recognition probability.

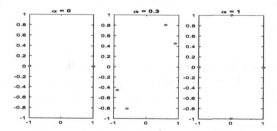

Fig. 3. The constellation fission function of BPSK

Considering both $F^\alpha(\cdot)$ and $T^\alpha(\cdot)$, with the increase of α, the two constellation points of BPSK gradually rotate clockwise and disperse. As α increases further, the boundaries between constellation points become blurred, and eventually the constellations overlap, as shown in Fig. 4. It is evident that transformed signals present a Gauss-like distribution after constellation aliasing. It can be concluded that the rotation, aliasing and Gaussian distribution of the constellation make it difficult to detect and identify modulation of the transformed signal, thus WFRFT is an effective anti-detection signal encryption method.

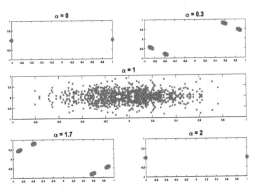

Fig. 4. WFRFT constellations with different α

In the SATCOM system, WFRFT can be used to replace FFT/IFFT in traditional TDCS in order to gain better anti-eavesdropping capacity. In the wake of the adjustment to transform coefficients, the spatial distribution of signal is changed and the parameters are difficult to be scanned. In a nutshell, WFRFT increases anti-interception capacity, while maintaining the complexity of the system.

3   WFRFT-TDCS System

TDCS is a spread spectrum technology in transform domain. In general, it is used to resist interception, avoid interference and weaken the influence of multipath. In response to the issue of low security and easy interception in the SATCOM system, this paper presents an approach to low interception communication based on WFRFT-TDCS.

3.1   System Model

Figure 5 shows the block diagram of the system in the satellite channel. On the transmitter side, the entire spectrum band is divided into N spectrum bins and a spectrum occupancy vector A^{TX} is used to indicate the status of spectrum bins. Then, A^{TX} is multiplied by a M-valued complex pseudo-random phase sequence P, to get a new vector B^{TX}. In a traditional TDCS architecture, the resultant spectral vector B^{TX} performs inverse fast Fourier transform (IFFT) operation to produce a time-domain signal b^{TX}. b^{TX} is modulated with the original data using cyclic code shift keying (CCSK) scheme. The transmitted signal is given by

$$S_{\text{IFFT}}(n) = \frac{1}{\sqrt{N}} \sum_{k=0}^{N-1} B^{\text{TX}}(k) \exp\left(j\frac{2\pi(k+S_i)n}{N}\right)$$ (6)

where S_i is the data symbol arriving at the input of CCSK modulator during the i time slot.

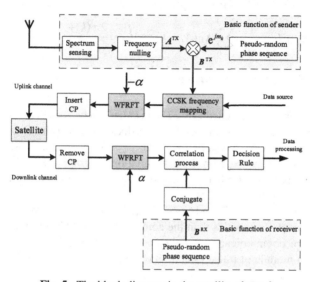

Fig. 5. The block diagram in the satellite channel

In this paper, orthogonal frequency division multiplexing (OFDM) architecture is adopted in the TDCS. Firstly, CCSK frequency-domain mapping is performed on the BF **B** according to circular shift theorem in time domain. Then WFRFT is used to replace IFFT in traditional TDCS. After that, the cyclic prefix is added to eliminate the impacts of multipath fading channel. Combined with the constellation fission function $T^\alpha(\cdot)$ and the Gauss-like function $F^\alpha(\cdot)$, the transmitted signal can be written as

$$S_{\text{F - WFRFT}}(n) = \omega_0(\alpha)B(n)\exp\left(j\frac{2\pi S_i n}{N}\right)$$

$$+ \omega_1(\alpha)\frac{1}{\sqrt{N}}\sum_{k=0}^{N-1} B(k)\exp\left(j\frac{2\pi(k+S_i)n}{N}\right)$$

$$+ \omega_2(\alpha)B(N-n)\exp\left(j\frac{2\pi S_i n}{N}\right)$$

$$+ \omega_3(\alpha)\frac{1}{\sqrt{N}}\sum_{k=0}^{N-1} X_0(N-k)\exp\left(-j\frac{2\pi(k+S_i)n}{N}\right)$$

$$= T^\alpha(\boldsymbol{B}) \exp\left(j\frac{2\pi S_i n}{N}\right) + F^\alpha(\boldsymbol{B}) \exp\left(j\frac{2\pi S_i n}{N}\right) \tag{7}$$

By comparison, the BF $\boldsymbol{B}$ is first processed by WFRFT, and then CCSK modulation is operated in time domain. The transmitted signal can be expressed as

$$S_{\text{T - WFRFT}}(n) = \omega_0(\alpha) B(n + S_i) + \omega_2(\alpha) B(N - n - S_i)$$

$$+ \omega_1(\alpha) \frac{1}{\sqrt{N}} \sum_{k=0}^{N-1} B(k) \exp\left(j\frac{2\pi(k + S_i)n}{N}\right)$$

$$+ \omega_3(\alpha) \frac{1}{\sqrt{N}} \sum_{k=0}^{N-1} X_0(N - k) \exp\left(-j\frac{2\pi(k + S_i)n}{N}\right)$$

$$= T^\alpha\left(\boldsymbol{B}_{S_i}\right) + F^\alpha(\boldsymbol{B}) \exp\left(j\frac{2\pi S_i n}{N}\right) \tag{8}$$

According to above-mentioned Eqs. (7) and (8), it can be seen that both the time and frequency signal processing methods of CCSK does not affect the Gauss-like characteristics of $F^\alpha(\cdot)$. However, constellation points of $T^\alpha(\cdot)$ based on CCSK time-domain modulation are relatively fixed on the constellation diagram, while the constellation points of $T^\alpha(\cdot)$ based on CCSK frequency-domain mapping are randomly distributed and affected by data symbol S_i.

The combination of WFRFT and CCSK frequency-domain mapping has better constellation splitting characteristics than the combination with CCSK time-domain modulation. In secure communication, as shown in Fig. 6. WFRFT is recommended to utilize as the last module of transmission, in order to make full use of its anti-scanning characteristics.

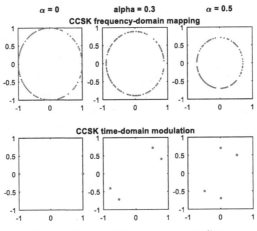

Fig. 6. The constellation points of $T^\alpha(\cdot)$

3.2 Design of Receiving Basis Function

The spectrum mismatch leads to BF inconsistency, and inevitably has negative effects on TDCS: (1) the signal is transmitted in mismatched frequency bands, resulting in the performance penalties in demodulation; (2) the bandwidth reduction of BF causes the interruption of communication. In CR networks, the public control channels are established to transmit the BF among transmitters. Nevertheless, it will increase complexity of SATCOM system and do not meet the security requirements. To solve the problem, the arithmetic of EGC is used to design the receiving BF.

From TDCS correlation receiving process point of view, the received signal energy is equal to the square of the product of the transmitter and the receiver's amplitude spectrums and phase spectrums. This is similar to the principle of the diversity combining in fading channels.

Under the following three known conditions: (1) the transmitter's amplitude spectrum is unknown; (2) the pseudo-random phase sequence is known; (3) the received signal contains pseudo-random phase information, so the pseudo-random phase sequence P can be used as receivers' BF, i.e. $B^{RX} = P$. The design of the receiver's BF can be realized by EGC principle.

4 Simulation Results

In this section, simulation results are provided to evaluate the performance of WFRFT-TDCS scheme.

4.1 Data Transmission Performance

With the expansion of high frequency band in SATCOM, electromagnetic interference of terrestrial communication systems can be effectively avoided. Hence, it can be considered that the ratio of BF spectrum available at both transmitter and receiver is relatively high for Ka, Ku and other higher frequencies. In this simulation, the available ratio ψ of BF spectrum at the transmitter is 60%, 80%, 95% and 100% respectively.

Because the receiver's BF is designed in the way of EGC, the bit error rate (BER) is not affected by the mismatched proportion of transceiver spectrum. As can be seen from Fig. 7, with the increase of spectrum availability, BER performance is also improved. SNR loss is less than 1 dB when ψ is more than 80%.

4.2 Anti-interception Performance

The BER curves of authorized and unauthorized receivers are compared to verify the effectiveness of WFRFT-TDCS technology. The parameters of the receivers are shown in Table 1.

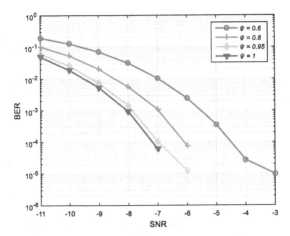

Fig. 7. BER performance with different ψ

Figure 8 shows that the error rates of unauthorized receivers NO.2 and NO.3 are still high, even if most of the transmission parameters are obtained in high SNR. This means that unauthorized receivers are difficult to obtain reliable covert information. We can come to conclusion that the system proposed in this paper has high anti-scanning capacity. The BER curves of unauthorized receivers NO. 1 and NO. 4 indicate that if the system adopts TDCS or WFRFT independently, unauthorized receivers can decode signals by traversal method. Therefore, in order to improve the security and guarantee the anti-interception ability of SATCOM system, it is better to combine two anti-eavesdropping means, TDCS and WFRFT, which is more difficult for unauthorized receivers to obtain all the system parameters.

Table 1. Parameters of authorized and unauthorized receivers

Parameter	Authorized receiver	Unauthorized receivers NO. 1	Unauthorized receivers NO. 2	Unauthorized receivers NO. 3	Unauthorized receivers NO. 4
Spectrum occupancy A^{RX}	EGC	Random number generation	EGC	EGC	EGC
Pseudo-random phase sequence P	Known	Known	Unknown	Known	Known
Transform techniques	WFRFT	WFRFT	WFRFT	FFT	WFRFT
WFRFT coefficient α	0.3	0.3	0.3	0.3	0.2

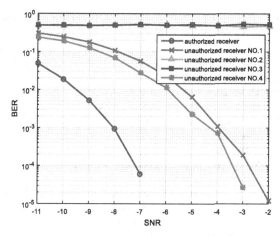

Fig. 8. BER performance of authorized and unauthorized receivers

5 Conclusion

In order to improve the safety and reliability of SATCOM, an anti-interception technology based on WFRFT-TDCS is proposed. WFRFT is used to replace FFT/IFFT and CCSK frequency-domain mapping is used to replace time-domain modulation in TDCS. Besides, the receiving BF is designed according to the principle of EGC. Final simulations demonstrate that WFRFT-TDCS anti-eavesdropping technology can be applied to satellite transceiver of spectrum mismatch, as well as maintain system complexity and improve anti-scanning performance.

SATCOM transmission reliability can be further improved by designing complex transceiver BFs, such as exchanging spectrum information between the transceiver and receiver. The adjustment strategy of WFRFT parameters will be further studied to better adapt to time-varying satellite channels.

References

1. German, E.H.: Transform domain signal processing study final report. Wright Patterson (1988)
2. Radcliffe, R.A., Gerald, C.G.: Design and simulation of a transform domain communication system. In: IEEE MILCOM, Monterey, pp. 586–589 (1997)
3. Xie, T.C., Da, X.Y., Zhu, Z.Y., et al.: Estimation of basis functions sequence for transform domain communication systems based on CCSK modulation. J. Syst. Simul. **26**(8), 1713–1717 (2014)
4. Lu, W.D., Gong, Y., Ting, S.H., et al.: Cooperative OFDM relaying for opportunistic spectrum sharing: protocol design and resource allocation. IEEE Trans. Wireless Commun. **11**(6), 2126–2135 (2012)
5. Xie, T.C., Da, X.Y., Zhu, Z.Y., et al.: Basis function design for transform domain communication system in the present of spectral mismatches. J. Jilin Univ. (Eng. Technol. Ed.) **44**(6), 1825–1830 (2014)
6. Liu, J.Y., Su, Y.T.: Performance analysis of transform domain communication systems in the presence of spectral mismatches. In: IEEE MILCOM, Orlando, pp. 1–5 (2007)

7. Mei, L., Sha, X.J., Zhang, N.T.: The approach to carrier scheme convergence based on 4-weighted fractional Fourier transform. IEEE Commun. Lett. **14**(6), 503–505 (2010)
8. Liang, Y., Da, X.Y., Xu, R.Y., et al.: Research on constellation-splitting criterion in multiple parameters WFRFT modulations. IEEE Access **6**, 34354–34364 (2018)
9. Zhai, D., Da, X.Y., Liang, Y., et al.: Satellite anti-interception communication system with WFRFT and MIMO. In: 2018 10th International Conference on Communication Software and Networks, Chengdu, pp. 305–310 (2018)
10. Liang, Y., Da, X.Y.: Analysis and implementation of constellation precoding system based on multiple parameters weighted-type fractional Fourier transform. J. Electron. Inform. Technol. **40**(4), 825–831 (2018)
11. Da, X.Y., Liang, Y., Hu, H., et al.: Embedding WFRFT signals into TDCS for secure communications. IEEE Access **6**, 54938–54951 (2018)
12. Wang, S., Da, X.Y., Zhu, Z.Y., et al.: Secure transmission for TDCS using 4-WFRFT and noise insertion. J. Sichuan Univ. (Eng. Sci. Ed.) **48**(3), 142–147 (2016)
13. Da, X.Y., Lian, C.: Method of weighted Fourier transform domain communication. Syst. Eng. Electron. **37**(12), 2853–2859 (2015)
14. Xu, R.Y., Da, X.Y., Liang, Y., et al.: Modified WFRFT-based transform domain communication system incorporating with spectrum mismatching. K SII Trans. Internet Inf. Syst. **12**(10), 4797–4813 (2018)

Wireless Signal Recognition Based on Deep Learning for LEO Constellation Satellite

Xin Zhou[✉], Yichen Xiao, Mingming Hu, and Lixiang Liu

Science and Technology on Integrated Information System Laboratory, Institute of Software
Chinese Academy of Sciences, Beijing 100190, China
zhou_200391@163.com

Abstract. In view of the increasing on-board processing capacity, this paper investigates the possibility of the communication reconnaissance on LEO constellation satellite platforms, and proposes a wireless signal recognition algorithm based on deep learning. The proposed algorithm visualizes the wireless signal as a picture based on the basic digital signal processing, as a result, the signal recognition problem is subtly transferred to an object detection problem recurring in the field of Computer Vision (CV). Then, it co-opts deep learning models in CV field in order to realize the end-to-end signal recognition and improve the performance. Validating results on the field-collected signal dataset with 12 types and 4740 samples show that, the algorithm can effectively identify the waveform types and time/frequency coordinates of communication signals with the precision 89%, which is 40% higher than traditional algorithms.

Keywords: Wireless signal recognition · Radio frequency deep learning · Convolutional neural network · LEO constellation satellite

1 Introduction

Influenced by the earth curvature and ground obstacles, UHF and higher frequency electromagnetic signals decay rapidly when they propagate on the ground, which limits the ground-based communication reconnaissance to a small range. However, the non-directional radiation signal emitted by the ground or low altitude communication target may be captured and recognized by the LEO satellite, owing to the moderate distance and the non-occlusion link between them. Furthermore, using the global coverage of the LEO constellation, space-based communication reconnaissance can be realized in the whole world. This can expand the area of communication reconnaissance and reduce the cost, which is significant to the communication countermeasure and security.

Communication signal recognition is essentially a kind of "pattern recognition", so it may be combined with artificial intelligence (AI) technology to improve the performance. This topic has been investigated extensively since 1990s. As shown in Fig. 1, these kind of algorithms generally relay on the expert feature extraction, and classify signals based

This work is supported by the National Key R&D Program (2016YFB0501104) and Special Fund for National Defense Technology Innovation (18-163-11-ZT-003-027-01).

© Springer Nature Singapore Pte Ltd. 2020
Q. Yu (Ed.): SINC 2019, CCIS 1169, pp. 275–285, 2020.
https://doi.org/10.1007/978-981-15-3442-3_23

on the traditional machine learning (ML) such as Decision Tree, Naive Bayes, SVM or Neural Networks [1–4]. They can gain a respectable performance in Additive White Gaussian Channel (AWGN), however degrade sharply in fading channel ubiquitous in the real world.

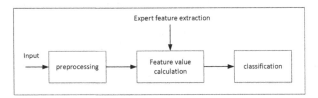

Fig. 1. Architecture of signal recognition based on traditional machine learning

Inspired by the tremendous success of deep learning (DL) in the field of Computer Vision (CV), some scholars began to use DL technology to improve the wireless signal recognition performance since 2016. O'shea proposed a 4-layer CNN (convolutional neural network) architecture to recognize the modulation type of wireless signal, directly feeding with the raw IQ samples [5], and demonstrated the superiority of the deep learning. Based on this, [6] used a hierarchical structure to improve the performance, [7, 8] investigated the LSTM (long-short term model) architecture and [9] investigated the ResNet architecture.

The above literatures belong to modulation recognition, while the following belongs to communication system recognition, which is more meaningful for communication reconnaissance. [10] used the neuro-fuzzy similarity measurement based on the signal power spectral density to distinguish signals. [11, 12] proposed a model similar as the O'shea CNN to identify the wireless technology, except that the input is replaced with spectrum density. [13, 14] adopted a similar network model, but replaced the input with the spectrum waterfall or Choi-Williams distribution transformation result.

This paper proposes a wireless signal recognition algorithm based on end-to-end deep learning. It does not follow the traditional idea of expert feature extraction plus automatic classification, but co-opts deep CNN in CV field [16–18] to automatically extract waveform features, and directly identifies the communication system types and time/frequency coordinate.

The rest paper is organized as follows. Section 2 analyzes the satellite-ground link loss of LEO, and investigates the possibility of the space-based communication signal recognition. Section 3 introduces the proposed algorithm. Section 4 shows the evaluation results, and Sect. 5 concludes the paper.

2   Link Loss Analysis of LEO SAT

LEO satellites refer to satellites with orbital altitudes between 200 km and 2000 km [15]. Thanks to the moderate distance and the non-occlusion link, the non-directional radiation signal emitted by the ground or low altitude communication target may be captured and recognized by the LEO satellite.

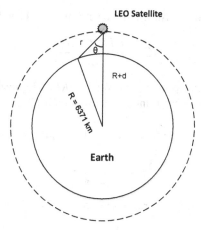

Fig. 2. Satellite-ground link of LEO

As shown in Fig. 2, assuming the orbital altitude d equals 600 km, and the scanning angle is $\pm30°$. According to cosine theorem shown in (1), the maximum distance r between the satellite and the ground target satisfies the Eq. (2). In this scenario, θ equals $30°$, R equals 6371 km, d equals 600 km, so r equals 704 km.

$$r^2 + (R+d)^2 - 2r(R+d)\cos\theta = R^2 \tag{1}$$

$$r = \frac{2(R+d)\cos\theta - \sqrt{(2(R+d)\cos\theta)^2 - 4(d^2 + 2Rd)}}{2} \tag{2}$$

According to the free space loss model shown in (3), the link loss L is about 157 dB when the center frequency F equals 2.5 GHz.

$$L = 32.4 + 20\log F + 20\log r \tag{3}$$

More scenarios are shown in Table 1. When the orbital altitude is 600 km, the S-band propagation attenuation between satellite and earth is about 160 dB.

Table 1. Propagation attenuation between satellite and earth

d/km	$\theta/°$	r/km	F/MHz	L/dB
600	0	600	2500	156.0
600	10	610	2500	156.1
600	20	642	2500	156.5
600	30	704	2500	157.3
600	40	811	2500	158.5
600	50	1006	2500	160.4
600	60	1450	2500	163.6

Generally speaking, the transmitter-receiver link model can be expressed as (4)

$$C = \frac{EIRP * G_r}{L_s} \qquad (4)$$

where C is the received carrier power in watts, $EIRP$ is Equivalent Isotropic Radiated Power in watts, G_r is the receiver antenna gain in multiples, and L_s is the propagation attenuation in multiples.

Divided by the receiver thermal noise N in both sides, we get

$$\frac{C}{N} = \frac{EIRP * G_r}{L_s * N} \qquad (5)$$

Denote $10 * \log(x)$ as $[x]$, and translating (5) into Log form, we get

$$\left[\frac{C}{N}\right] = [EIRP] + [G_r] - [L_s] - [N] \qquad (6)$$

Here, [C/N] is also called signal-to-noise ratio (SNR) which is the most important parameter to evaluate the received signal quality. According to the commutation theory, the thermal noise [N] satisfies the Eq. (7) when the temperature $T = 290$ K ≈ 17 °C.

$$[N](dBW) \approx -204 + 10\log B + NF \qquad (7)$$

where B is signal bandwidth in Hz and NF is the noise figure in dB.

Substitute (7) into (6), we get

$$SNR = [EIRP] + [G_r] - [L_s] - [B] - NF + 204 \qquad (8)$$

Table 2 shows the received signal quality in some scenarios. In engineering, the NF of the receiver on the satellite is about 2 dB, and the system should reserve a margin about 5 dB for the propagation attenuation L_s.

Table 2. Received signal quality evaluation

EIRP/dBW	Gr/dBi	Ls/dB	B/MHz	Margin/dB	SNR/dB
1	5	157.3	0.05	5	−1.3
1	5	157.3	2	5	−17.3
1	5	163.6	0.05	5	−7.6
1	5	163.6	2	5	−23.6
1	35	157.3	0.05	5	**28.7**
1	35	157.3	2	5	**12.7**
1	35	163.6	0.05	5	**22.4**
1	35	163.6	2	5	6.4

Based on the previous literation and our own test, the signal recognition algorithms can work well when SNR > 10 dB [9]. In other words, the recognition sensitivity is 10 dB. As shown in Table 2, if the antenna gain of the satellite is bigger than 35 dBi, the received SNR is higher than the recognition sensitivity and will works well for most narrow-bandwidth and mi-bandwidth signals. It's a good news for space-based communication reconnaissance!

3 The Proposed Algorithm

In this section, we will introduce the proposed wireless signal recognition algorithm based on deep learning as shown in Fig. 3. Its input is raw I/Q flow and the output is the signal types and corresponding time/frequency coordinates. Since there is a subtle visualization step in the algorithm, we call it *RadioImageDet*.

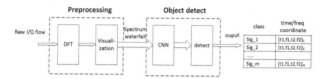

Fig. 3. Architecture of RadioImageDet algorithm

RadioImageDet algorithm consists of preprocessing module and object detect module. The purpose of preprocessing is to transform the original I/Q data into an expression more conducive to automatic feature extraction and recognition. Generally speaking, wireless communication signals show certain patterns in center frequency, bandwidth, spectrum density and their changing trends, which are more recognizable in frequency-domain. So, the preprocessing module in this algorithm takes Discrete Fourier Transform (DFT) to transform the input into frequency-domain expression as shown in (9), and then combines a group of successive DFT results into a spectrum waterfall map, in order to increase the time-domain information as shown in Fig. 4.

$$DFT : X[k] = \sum_{n=0}^{N-1} x[n]e^{-jk(2\pi/N)n}, \ k = 0, 1, \ldots, N-1$$

$$IDFT : x[n] = \frac{1}{N}\sum_{k=0}^{N-1} X[k]e^{jk(2\pi/N)n}, \ n = 0, 1, \ldots, N-1 \tag{9}$$

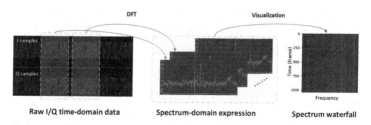

Fig. 4. Visualization preprocessing

The object detect module takes the spectrum waterfall map as the input, and automatically extracts the features through CNN. The following detect module transforms the abstract features into the final results such as signal type and time/frequency coordinates. The essence of this work is nearly the same as the object detection in CV. The only difference is that the input is a spectrum waterfall map obtained from DFT, not a picture in nature. Therefore, we co-opt the object detection models [17] in CV field to construct a wireless signal recognition neural network model, called RadioYOLO, as shown in Fig. 5 and Table 3.

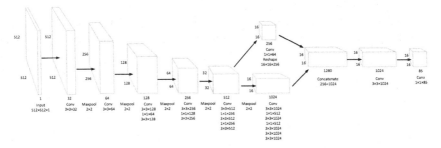

Fig. 5. RadioYOLO neural network model

Table 3. RadioYOLO model structure

Level	Input scalar	Scalar/step	Channels
conv1	512 × 512	3 × 3/1 × 1	32
maxpool1	512 × 512	2 × 2/2 × 2	–
conv2	256 × 256	3 × 3/1 × 1	64
maxpool2	256 × 256	2 × 2/2 × 2	–
conv3	128 × 128	3 × 3/1 × 1	128
conv4	128 × 128	1 × 1/1 × 1	64
conv5	128 × 128	3 × 3/1 × 1	128
maxpool3	128 × 128	2 × 2/2 × 2	–
conv6	64 × 64	3 × 3/1 × 1	256
conv7	64 × 64	1 × 1/1 × 1	128
conv8	64 × 64	3 × 3/1 × 1	256
maxpool4	64 × 64	2 × 2/2 × 2	–
conv9	32 × 32	3 × 3/1 × 1	512
conv10	32 × 32	1 × 1/1 × 1	256

(*continued*)

Table 3. (*continued*)

Level	Input scalar	Scalar/step	Channels
conv11	32 × 32	3 × 3/1 × 1	512
conv12	32 × 32	1 × 1/1 × 1	256
conv13	32 × 32	3 × 3/1 × 1	512
maxpool5	32 × 32	2 × 2/2 × 2	–
conv14	16 × 16	3 × 3/1 × 1	1 024
conv15	16 × 16	1 × 1/1 × 1	512
conv16	16 × 16	3 × 3/1 × 1	1 024
conv17	16 × 16	1 × 1/1 × 1	512
conv18	16 × 16	3 × 3/1 × 1	1 024
conv19	16 × 16	3 × 3/1 × 1	1 024
conv20	16 × 16	3 × 3/1 × 1	1 024
conv21	32 × 32	1 × 1/1 × 1	64
reshape1	32 × 32	2 × 2	256
concatenate1	16 × 16	–	1 280
conv22	16 × 16	3 × 3/1 × 1	1 024
conv23	16 × 16	1 × 1/1 × 1	85

RadioYOLO predicts the bounding boxes based on some hand-picked priors named *anchor*, which is an important hyper-parameter for the model. Five anchors are selected according to the common characteristics of radio signals, expressed as (width, height):

(1) (0.015, 0.9): for Narrowband Continuous Signal
(2) (0.5, 0.9): for Broadband Continuous Signal
(3) (0.025, 0.05): for Narrowband Instantaneous Signal
(4) (0.469, 0.01): for Broadband Instantaneous Signal
(5) (0.469, 0.1): for Broadband Instantaneous Signal

4 Experimental Evaluation

This research trains and validates the algorithm under the framework of Tensorflow, and evaluates the performance on field collected dataset. The comparison algorithms include the traditional NFSC (Neuro-Fuzzy Signal Classifier) algorithm [10] and the WII (multi-label Wireless Interference Identification) algorithm [12].

All data used in this paper is captured from the real world through a self-developed radio frequency acquisition test-bed based on Gnuradio and Universal Software Radio Platform (USRP), as shown in Fig. 6.

Computer   **USRP**
(with GnuRadio software)

Fig. 6. Radio frequency acquisition test-bed

The target signals cover 12 kinds of common 2G/3G/4G mobile communication downlink signals. 650 GB raw I/Q signal data with different center frequencies and the same sampling rate (40Msps) is collected from seven locations scattered at Beijing and Tianjin. These raw I/Q data are divided into fragments, which contain about 2 million I/Q samples, and transformed into waterfall map patterns with size 512 * 512. The resulting dataset has 4740 items, with 3792 for training and 948 for testing.

Data annotation is done manually according to the frequency allocation table. Table 4 shows the 2G/3G/4G cellular downlink frequency distribution of the three major operators in Beijing, which is attained through the relevant standards and actual measurement.

Table 4. 2G/3G/4G cellular downlink frequency distribution in Beijing

Frequency/MHz	Signal name
870–885	Tele2G_DL_CDMA
935–955	Mobi2G_DL_GSM900
955–960	Uni2G_DL_GSM900
1 805–1 820	Mobi2G_DL_GSM1800
1 840–1 860	Uni4G_DL_FDD1800
1 860–1 875	Tele4G_DL_FDD1800
1 885–1 905	Mobi4G_TDLTE_20M
1 905–1 915	Mobi4G_TDLTE_10M
2 010–2 025	Mobi3G_TDSCDMA
2 110–2 130	Tele4G_DL_FDD2100
2 130–2 135	Uni3G_DL_WCDMA
2 135–2 140	Uni3G_DL_WCDMA
2 140–2 155	Uni4G_DL_FDD2100
2 575–2 595	Mobi4G_TDLTE_20M

(continued)

Table 4. (*continued*)

Frequency/MHz	Signal name
2 595–2 615	Mobi4G_TDLTE_20M
2 615–2 635	Mobi4G_TDLTE_20M

Figure 7 shows the visual results of the proposed algorithm. It automatically marks the signal type, location and confidence on the original map, according to the numerical output. It can be seen that, the characteristics of some signals on the spectrum waterfall map are very obvious, so they can be easily distinguished by human and the proposed algorithm. This is the basic premise and motivation for our research.

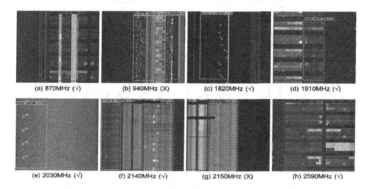

Fig. 7. Visual results of RadioImageDet algorithm

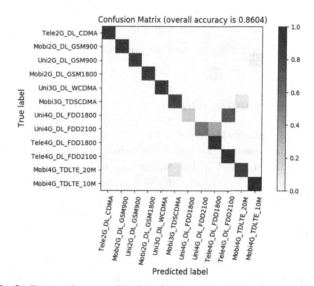

Fig. 8. Test result - normalized confusion matrix (Color figure online)

Figure 8 is the normalized confusion matrix of the test results of this algorithm. The vertical axis represents the real label, the horizontal axis represents the prediction label, and the depth of the grid color represents the magnitude of the value. Generally speaking, darker color of the diagonal matrix means better performance, and the other grids is on the contrary. The results show that the proposed algorithm performs well for most of signals with the overall accuracy 86.04%.

As comparison, WII and NFSC are also evaluated on the same raw dataset. The former uses deep learning method with a simpler CNN model, while the latter uses traditional method based on similarity measurement. We choose *precision* as the metric, which is defined as the ratio of the actual positive count to the predicted positive count for a specific type. As shown in Fig. 9, the average precision reaches 89% for the proposed algorithm, while 53.8% for WII and 49.3% for NFSC.

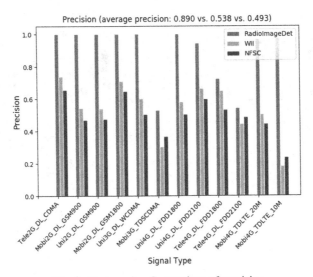

Fig. 9. Test result – Comparison of precision

5   Conclusion and Future Work

The development of LEO constellation network brings new opportunities for global communication reconnaissance. This paper firstly analyzed the satellite-ground link loss of LEO, and found that the SNR will be higher than 10 dB in some specific scenarios, which means that the space-based communication signal recognition is entirely feasible. Then, we dug into the wireless signal recognition, and proposes an algorithm based on deep learning. Experimental results show that the precision of the proposed algorithm reaches 89%, which is 40% higher than traditional method and 35% higher than simple CNN method.

This research is only a preliminary exploration of the space-based communication signal recognition based on deep learning. In the future, we will investigate the pre-processing methods which give more prominence to the waveform features, the neural network models which are more suitable for wireless signal recognition, and the effects of satellite-ground fading channel.

References

1. Chen, J., Kuo, Y., Li, J., Fu, F.: Review of automatic communication signals recognition (Chinese with English abstract). J. Circuits Syst. **10**(5), 102–109 (2005)
2. Wong, M.L.D., Nandi, A.K.: Automatic digital modulation recognition using spectral and statistical features with multi-layer perceptron. In: Signal Processing and its Applications, vol. 2, pp. 390–393 (2001)
3. Arulampalam, G., Ramakonar, V., et al.: Classification of digital modulated schemes using neural networks. In: International Symposium on Signal Processing and its Applications, vol. 2, pp. 649–652 (1999)
4. Triantafyllakis, K., Surligas, M., Vardakis, G.: Phasma: an automatic modulation classification system based on random forest. In: IEEE International Symposium on Dynamic Spectrum Access Networks, vol. 1, pp. 1–3 (2017)
5. O'Shea, T.J., Corgan, J., Clancy, T.C.: Convolutional radio modulation recognition networks. In: Jayne, C., Iliadis, L. (eds.) EANN 2016. CCIS, vol. 629, pp. 213–226. Springer, Cham (2016). https://doi.org/10.1007/978-3-319-44188-7_16
6. Karra, K., Kuzdeba, S., Petersen, J.: Modulation recognition using hierarchical deep neural networks. In: IEEE International Symposium on Dynamic Spectrum Access Networks, vol. 1, pp. 1–3 (2017)
7. West, N., O'Shea, T.: Deep architectures for modulation recognition. In: IEEE International Symposium on Dynamic Spectrum Access Networks, vol. 1, pp. 1–6 (2017)
8. West, N., Harwell, K., McCall, B.: DFT signal detection and channelization with a deep neural network modulation classifier. In: IEEE International Symposium on Dynamic Spectrum Access Networks, vol. 1, pp. 1–3 (2017)
9. O'Shea, T., Roy, T., Clancy, T.C.: Over the air deep learning based radio signal classification. IEEE J. Sel. Top. Sign. Proces. **12**(1), 168–179 (2017)
10. Ahmad, K., Shresta, G., Meier, U., et al.: Neuro-fuzzy signal classifier (NFSC) for standard wireless technologies. In: International Symposium on Wireless Communication Systems, vol. 1, pp. 616–620 (2010)
11. Schmidt, M., Block, D., Meier, U.: Wireless interference identification with convolutional neural networks. In: IEEE 15th International Conference on Industrial Informatics, vol. 1, pp. 180–185 (2017)
12. Grunau, S., Block, D., Meier, U.: Multi-label wireless interference identification with convolutional neural networks (2018)
13. Bitar, N., Muhammad, S., Refei, H.H.: Wireless technology identification using deep convolutional neural networks. In: IEEE 28th Annual International Symposium on Personal, Indoor, and Mobile Radio Communications (PIMRC), pp. 1–6 (2017)
14. Zhang, M., Diao, M., Guo, L.: Convolutional neural networks for automatic cognitive radio waveform recognition. IEEE Access **5**(1), 11074–11082 (2017)
15. Pratt, T., Bostian, C., Allnutt, J.: Satellite Communications, 2nd edn. Publishing House of Electronics Industry, Beijing (2005)
16. Ren, S., He, K., Girshick, R., Sun, J.: Faster R-CNN: towards real-time object detection with region proposal networks. IEEE Trans. Pattern Anal. Mach. Intell. **39**(6), 1137–1149 (2017)
17. Redmon, J., Farhadi, A.: YOLO9000: better, faster, stronger. In: IEEE Conference on Computer Vision and Pattern Recognition, vol. 1, pp. 6517–6525 (2017)
18. He, K., Zhang, X., Ren, S., Sun, J.: Deep residual learning for image recognition. In: IEEE Conference on Computer Vision and Pattern Recognition, vol. 1, pp. 770–778 (2016)

Author Index

Printed in the United States
By Bookmasters